HARNESS THE SUN

HARNESS THE SUN

AMERICA'S QUEST FOR A SOLAR-POWERED FUTURE

PHILIP WARBURG

BEACON PRESS, BOSTON

Beacon Press
Boston, Massachusetts
www.beacon.org

Beacon Press books
are published under the auspices of
the Unitarian Universalist Association of Congregations.

18 17 16 15   8 7 6 5 4 3 2 1

This book is printed on acid-free paper that meets the uncoated paper
ANSI/NISO specifications for permanence as revised in 1992.

Text design by Wilsted & Taylor Publishing Services
Cover photo: Arizona State University, Tempe. Photo by author.

3/2017

Library of Congress Cataloging-in-Publication Data

Warburg, Philip.
Harness the sun : America's quest for a solar-powered future / Philip Warburg.
pages cm
Includes bibliographical references and index.
ISBN 978-0-8070-3376-0 (hardcover : alk. paper)
ISBN 978-0-8070-3377-7 (ebook)
1. Solar energy—United States. I. Title. II. Title: America's quest
for a solar-powered future.
TJ809.95.W37 2015
333.792'30973—dc23
2015003724

CONTENTS

Note on Terminology

IN THE PAGES THAT follow, I describe solar installations of highly diverse scale, from small arrays serving individual households to giant solar farms that send huge infusions of power to the grid. Most of these systems rely on the *photovoltaic effect*, which uses the energy in particles of sunlight, or photons, to create electricity. A *photovoltaic (PV) panel*, also called a *module*, consists of multiple PV *cells*. The direct (DC) current generated by these cells must be converted to alternating (AC) current before it can be used by our electric system.

In some larger solar installations, the sun's heat rather than its light is captured and is then typically used to create steam for a conventional electricity-generating turbine. This latter approach, called *concentrating solar power* (CSP), is described in chapter 7.

Individual PV panels have a few hundred watts of *installed capacity*—the maximum power that they can generate under *standard test conditions* (described more fully in chapter 8). Household-scale solar systems typically have a few *kilowatts* (1,000 watts) of installed capacity. Solar installations on commercial and public buildings commonly have hundreds of kilowatts of installed capacity, and larger commercial and utility-scale solar plants can have multiple *megawatts* (1 million watts) of installed capacity. When quantifying solar power at the state or national level, the *gigawatt* (1 billion watts) is often used as the unit of measurement. The actual amount of electricity produced by solar facilities (measured in *kilowatt-hours, megawatt-hours,* or *gigawatt-hours*) depends on numerous factors including weather, temperature, and the particular technologies employed.

The Solar Energy Industries Association (SEIA) estimates that a megawatt of installed photovoltaic (PV) capacity can supply enough

Fig. 1. *Average Number of Homes Powered by a Megawatt of Solar Photovoltaics (25 States & District of Columbia)*

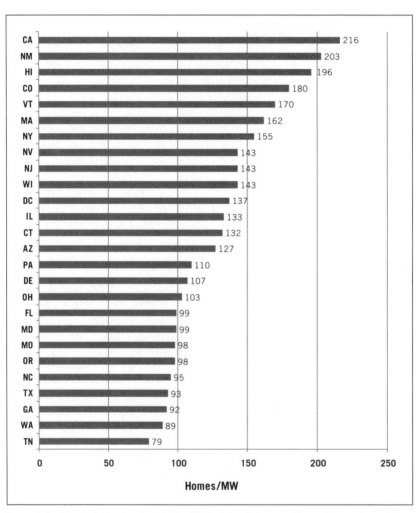

electricity for 164 American homes. The actual number of homes served by a megawatt of solar power varies quite widely from state to state, however, depending on differences in climate and household electricity use. (See fig. 1.) While there are many non-residential uses of electricity, the individual household is a favored metric for solar electric output because it is relatively immediate to most of our lives, and is therefore easy to understand.

Introduction

SOLAR ENERGY'S TIME HAS COME. Just a few years ago, as I was putting the finishing touches on a book about wind power, friends asked if my next would be about solar.[1] I dismissed the idea at the time. If I ever wrote a book about solar, I told them, I'd have to call it "Dim Sun."

How things have changed! On my computer screen, I marvel at the steady tick of kilowatt-hours produced by our home's rooftop solar array. It generates about three-quarters of the electricity we need to run our appliances, light up our rooms, and keep a hybrid electric vehicle fully charged. Every year our solar panels spare the globe about 3.6 tons of CO_2 emissions from the coal- and gas-fired plants that still supply most of New England's power.

At last we have entered an era when solar energy is not just an environmental virtue but is also a boon to the economy. More than 640,000 Americans have already installed solar power on their homes and businesses,[2] and almost 174,000 US workers have found jobs in the solar industry.[3] Those jobs will grow in the years ahead, as solar energy reaches millions of American homes, businesses, farms, public buildings, and the portfolios of power companies large and small. Over a century has passed since Albert Einstein identified the sun's photoelectric effect in 1905—a discovery that later earned him the Nobel Prize for Physics. I can only imagine how delighted he would be by our progress, however belated, in harnessing this formidable source of power.

It seemed that a new solar era was dawning in the years following the 1973 Arab oil embargo. President Jimmy Carter proclaimed a solar revolution, calling for a fifth of America's energy to come from the sun by the year 2000. The first federally designated Sun Day was held in 1978, and the government's newly created Solar Energy Research Institute pressed ahead with a bold agenda for solar R&D. With strong bipartisan congressional support, Carter authorized tax incentives that created a burst of solar energy investment. While a few power plants were built using mirrors to concentrate the sun's heat for electricity production, photovoltaic technology—converting sunlight to solar power—remained a rare luxury, primarily used on satellites where the need for electricity in orbital space justified the exorbitant cost. Solar water heaters, much cheaper and easier to build, were of greater practical value here on earth, and Carter—amid much fanfare—had one such system installed at the White House.

Then came the bust. In the 1980s Ronald Reagan purged federal support for renewable energy and ordered Carter's solar water-heating panels stripped from the White House roof. Unable to compete with cheap western coal and deregulated natural gas, solar energy's fate seemed to be sealed.

Yet today Americans are investing more money, and greater hope, in solar energy than ever before. Though solar energy has many applications, producing electric power from the sun has taken center stage as US and foreign companies vie for a share of this rapidly expanding market.[4] Along with shrinking photovoltaic (PV) manufacturing and installation costs, favorable policies at the state and federal levels have created whole new cohorts of solar stakeholders ranging from homeowners and businesses to electric utilities and financial institutions. During 2014, a new solar system was completed at the rate of one every 2.5 minutes, reflecting an overall investment of $17.8 billion that year.[5] Even more impressive is solar power's share of US investment in new electrical generation. In 2014, solar accounted for 32 percent of all new generating capacity, outpaced only by natural gas power plants, which captured 42 percent of new capacity. Wind came in third, at 23 percent, while investment in new coal plants was negligible.[6]

Much more than an echo of our earlier flirtation with renewable energy, solar power is part of a sea change that is sweeping through university labs, policy think tanks, and corporate boardrooms, where recognition is growing that a fundamental shift away from carbon-based fuels is imminent and inevitable. The devastating impacts of climate change may be too seldom acknowledged by our political leaders, but the overwhelming scientific evidence has persuaded technology innovators as well as broad segments of the public to embrace a major push toward renewable energy.

Solar electricity supplies less than half a percent of America's power today, yet it's no idle dream to foresee a quarter or more of our electricity coming from the sun in a few decades' time.[7] The National Renewable Energy Laboratory (NREL) has estimated America's solar potential at more than a hundred times our total electricity consumption, with rooftop installations alone capable of supplying over a fifth of our power needs—a contribution that could grow as solar technology improves.[8] The same government lab has projected that solar and wind power could provide almost half of America's electricity by 2050, using technology that is commercially available today. These technologies, together with hydropower, biomass-fueled power plants, and other renewable sources of electricity, could generate 80 percent of our mid-century electric output, NREL predicts.[9] Imagine the impact that would have on America's carbon footprint!

Despite all this promise, America has some catching up to do. We may have far greater land resources and sunnier weather, on the whole, than much of Europe, yet we lag far behind many European nations in our per capita use of solar power. Germany leads the world in both per capita and total use of solar energy, despite its notoriously gray climate. A highly publicized 100,000 Roofs Program got the ball rolling back in 1998, and vigorous subsidies raised that country's solar power to 436 watts per person by 2013—more than 11 times our own reliance on the sun, which stood at 38 watts per capita that year. Italy earned second place with 294 watts of solar power per person; next came Belgium, Greece, and the Czech Republic. In all, eighteen European countries delivered higher per capita solar use than the United States. (See fig. 2.)

Fig. 2. Solar Photovoltaic Cumulative Installed Capacity in Top 25 Countries (2013)

COUNTRY	INSTALLED CAPACITY (MEGAWATTS)	RANK	PER CAPITA CAPACITY (WATTS)
Germany*	35,715	1	436
China	18,300	2	13.5
Italy*	17,928	3	294
Japan	13,643	4	107
US	12,022	5	38
Spain	5,340	6	116
France*	4,673	7	71
UK*	3,375	8	53
Australia	3,255	9	145
Belgium*	2,983	10	268
Greece*	2,579	11	229
India	2,319	12	1.9
Czech Rep.*	2,175	13	207
S. Korea	1,467	14	29.9
Canada	1,210	15	35
Romania*	1,151	16	54
Bulgaria*	1,020	17	140
Switzerland	737	18	92
Thailand	704	19	10.4
Netherlands*	665	20	40
Ukraine*	616	21	14
Austria*	613	22	72
Denmark*	548	23	98
Slovakia*	524	24	97
Israel	420	25	54

Sources: For countries with asterisks, totals for cumulative installed capacity and per capita capacity were provided, with permission, by European Photovoltaic Industry Association (EPIA) (*Global Market Outlook for Photovoltaics 2014–2018*, fig. 6). For other countries, totals for cumulative installed capacity were provided, with permission, by International Energy Agency (IEA) (*PVPS Report—Snapshot of Global PV 1992–2013*, table 2). For countries whose cumulative capacity was reported by IEA, per capita capacity was derived by Philip Warburg from country population estimates in *CIA World Factbook*, July 2014.

But American solar power is growing fast. By the end of 2014, we had enough solar electric-generating capacity to supply all the needs of four million American households.[10] The solar power installed during that year alone amounted to nearly half of all previously installed capacity.[11] If we can maintain that momentum in the coming years, a robust, solar-powered future is within our reach.

—∞—

My personal solar journey began in our century-old home, just a short trolley ride from downtown Boston. I have spent most of my career as an environmental watchdog, railing against corporate polluters and advocating for cleaner energy. My wife, Tamar, is an architect who has designed many green buildings for others. As the price of solar panels spiraled downward in recent years, we decided it was time to make renewable energy a part of our personal as well as our professional lives. Our new solar array does just that.

But solar power's promise extends far beyond environmentally minded homeowners who are looking for ways to reduce their families' outsized carbon footprints. The dollars-and-cents reality today is that solar power is a good investment. Why else would business titan Warren Buffett buy up California solar farms worth several billion dollars? Why would Google be investing $280 million in a fund to finance home solar systems and hundreds of millions more in utility-scale solar plants that feed their power into the grid? And what other motivation would drive Walmart to place rooftop solar arrays on 250 of its properties, with plans to reach a thousand of its stores and warehouses by 2020?

Installing solar panels on our home was the beginning of an exploration that took me to football arenas, big-box stores, industrial warehouses, and college campuses where the sun's energy is being tapped. I also visited cities, towns, and counties whose political and business leaders are pioneers in advancing solar energy. Some of these communities defy easy stereotypes about renewable energy as the domain of enlightened liberals and well-to-do suburbanites. In

Lancaster, California, ultra-right-wing mayor Rex Parris has set his formerly crime-ridden city on a course to becoming what he brashly calls "the solar capital of the world," where solar power is mandated in all new residential neighborhoods, PV panels have been installed on nearly all public buildings, and utility-scale solar power plants are quickly going up on vacant land within the city's borders. Lancaster's demographics contrast sharply with upscale, super-progressive Marin County, where clean energy maverick Dawn Weisz has waged a successful campaign to create a community-based alternative to buying up power from utility giant PG&E. Here in my home state of Massachusetts, I saw how political savvy has fused with economic acumen to make the working-class city of New Bedford, once a center of whale oil production, a state leader in solar energy development.

America's brownfields—our abandoned and under-used industrial sites—are a huge, yet largely unrecognized solar resource. My travels took me to several of these sites: a garbage mountain overlooking lower Manhattan, a contaminated former factory site on Chicago's South Side, and a rocket propulsion test site east of Sacramento were among them. On these visits I met entrepreneurs and community leaders who have succeeded in turning gritty wastelands into renewable energy powerhouses. According to the US Environmental Protection Administration, harvesting the sun on brownfield properties like these could supply more than seven times the power needs of every household in the nation.

Appealing though it may be to capture the sun's energy as it falls onto our built environment, some of our greatest solar power opportunities lie outside our cities and towns, in open spaces where we can build solar farms that—at their peak—can match the output of coal- and gas-fired generating stations. By deploying solar technology across thousands of acres, we can also achieve economies of scale that are beyond the reach of smaller, scattered solar installations on our homes, commercial and public buildings, and urban brownfields. (See fig. 3.) About half of all solar power in America today comes from sprawling solar complexes like the ones I visited in the Arizona, California, and Nevada deserts. With proper access to the grid, these

Fig. 3. *Photovoltaic Systems—Price per Watt (in 2014 US$)*

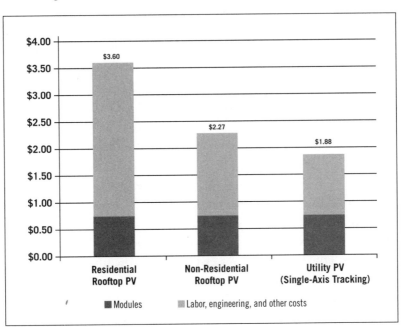

Source: SEIA/GTM Research, *U.S. Solar Market Insight Q3 2014.* http://www.greentechmedia.com /research/ussmi. For a description of the "bottom-up" methodology used in calculating these prices, see the Executive Summary of this report, 12, http://www.seia.org/.

renewable energy powerhouses can be vital suppliers of electricity to major metropolitan areas hundreds of miles away.

America's open spaces make for easy construction of large-scale solar power complexes, but I quickly learned that one person's undeveloped land is another's biodiversity treasure. Solar developers and conservationists are working hard—and often together—to ensure that utility-scale projects, when built, pose minimal harm to vulnerable earthbound species like the desert tortoise and the giant kangaroo rat. They are also grappling with the risks posed to birdlife by solar power towers, which use thousands of giant mirrors to convert the sun's heat to electricity-producing steam.

My travels also took me to a number of Native American communities, where I spoke with tribal officials and entrepreneurs about the solar initiatives that they are beginning to advance. Some tribal leaders see solar power as a way to build a more sustaining and respectful relationship to traditional lands while bringing much-needed income to their tribes. Others suspect that solar projects may simply be the latest attempt by non-Native peoples to exploit tribal energy resources. They remember the uranium mines that sickened so many of their people until they were finally banned a decade ago—on Navajo lands at least.[12] They are intimately aware of the hardships as well as the economic gains that have come from the coal mines and coal-fired power plants that they allowed to be built on their reservations. The jobs were—and still are—desperately needed, but at what cost to their traditional landscapes and their people's health?

Making the sun a major American energy resource isn't just about building new power plants; it's also about producing and responsibly handling all the equipment we need to harness the sun's energy. Along with taking the pulse of American solar manufacturing in the face of fierce competition from China, I investigated some of the knottier issues of solar waste management that will emerge in the years and decades ahead. Industry leaders will soon need to come up with a responsible way to deal with the billions of discarded panels and other solar hardware that will accumulate as we ramp up our use of solar power.

Of paramount importance are the state and federal policies that are now making it possible for solar energy to compete in a marketplace still dominated by carbon-based fuels and nuclear power. Are energy pundits exaggerating when they warn of a utility death spiral brought on by a shift to solar and other decentralized modes of power generation, storage, and use? Is there a new role for electric utilities in balancing these new energy resources with conventional means of electricity generation and distribution? In grappling with these questions, I sought out the wisdom of cool-headed analysts and the emboldened visions of industry iconoclasts.

America's shift toward solar power is one piece of an enormously challenging puzzle. The Intergovernmental Panel on Climate Change warns that we will need to reach 80 percent reliance on low-carbon fuel sources for our global electricity needs by 2050, and then move almost entirely away from the burning of fossil fuels for power generation by 2100, if we are to avert potentially catastrophic climate change.[13] Our glaciers are already melting; our oceans are warming; and extreme weather events are exacting a mounting toll on human communities, natural habitats, and vulnerable wildlife species. What we are witnessing is no more than a gentle preview of the havoc that awaits us if we ignore the overwhelming scientific consensus about rising greenhouse gas concentrations and their global impact. Left unchecked, the burning of fossil fuels will disrupt the lives of millions of people, raise the specter of war over food and water scarcity, and irreparably damage terrestrial and aquatic ecosystems.

As one of the biggest producers of the greenhouse gases that threaten our global environment, America has a duty to lead the way, rather than lag behind, in making these momentous shifts. Through the personal choices we make, the policies we adopt, the technical genius we apply, and the entrepreneurial spirit we engage, we can make a difference.

CHAPTER ONE

Our House, Your House

BACK IN 2003, WHEN my family moved into our new home in Newton, Massachusetts, it didn't occur to me that its roof would someday generate most of our household power needs plus the daily charge for an electric car. Solar panels at the time were prohibitively expensive, the chilly New England weather didn't strike me as a good match for solar energy, and the car market had barely entered the hybrid era.

Within a decade, our prospects for tapping the sun had brightened dramatically. The price of photovoltaic (PV) panels had dropped so low that, with the incentives offered by the federal government and states like our own, homeowners could install rooftop solar arrays and expect to recoup their investment in little more than half a dozen years. By the time I began shopping around for the right solar vendor in January 2013, hundreds of thousands of Americans had installed PV panels on their homes. Moreover, I learned why putting a PV array on a home in Massachusetts was not as incongruous as I had thought. Because the sun's light, rather than its heat, is what gets all those electrons flowing through the photosensitive cells of a PV panel, solar installations could benefit from the Bay State's relatively sunny climate, even on the coldest winter days. In fact, PV panels actually perform *best* in the bitter cold.[1] This, together with favorable state policies, has helped make Massachusetts the nation's fifth-ranked state for solar power development. (See fig. 4.)

Not quite sure where to start, I went online to see which companies had a track record of installing residential solar systems in eastern Massachusetts, and then invited a few to come by our home to make their pitch. Our first visitor was a disappointment. He arrived late,

Fig. 4. *Solar Electricity: Cumulative Installed Capacity in Top 30 States*

STATE	CUMULATIVE INSTALLED PV CAPACITY THROUGH 2014 (MWDC)	CUMULATIVE INSTALLED CSP CAPACITY THROUGH 2014 (MWAC)	SOLAR ELECTRIC CAPACITY PER CAPITA (WATTS/PERSON)
California	8,720.7	1,256	262.1
Arizona	1,786.0	283	315.6
New Jersey	1,451.1		163.5
North Carolina	953.2		97.8
Massachusetts	751.2		112.8
Nevada	725.0	64	286.3
Hawaii	440.5	7	321.3
Colorado	398.4		76.7
New York	396.9		20.2
Texas	330.0		12.6
New Mexico	324.6		155.7
Pennsylvania	244.8		19.2
Maryland	215.0		36.5
Georgia	161.2		16.2
Florida	159.2	75	12.1
Tennessee	130.0		20.1
Connecticut	118.6		33.0
Indiana	112.5		17.2
Missouri	111.1		18.4
Ohio	102.4		8.9
Oregon	84.5		21.7
Vermont	69.9		111.7
Louisiana	65.3		16.8
Delaware	60.7		66.2
Illinois	54.0		4.2
Washington	38.6		5.5
Minnesota	20.1		3.7
Wisconsin	19.8		
Utah	18.2		9.2
Virginia	11.2		2.2

Source: SEIA/GTM Research, *U.S. Solar Market Insight Report—2014 Year in Review.* PV capacity is measured in direct (DC) current, prior to conversion to alternating (AC) current. CSP capacity is measured in AC current. Solar electric capacity per capita is based on 2014 US Census Bureau estimates.

tracked mud onto our dining room rug, and brandished a Google Earth aerial photo that very clearly depicted the house next door. All of his pre-cooked calculations about roof orientation, the proposed placement of PV panels, and expected power output might have served our neighbors well. To us, they were irrelevant. He also boasted that our new solar array would raise our home's market value by more than we would pay for the system, an assertion that was as speculative as it was alluring.

Several days after this disappointing encounter, a sales representative from Sunlight Solar Energy came by our home. Though this Oregon-based company has only a modest presence in Massachusetts, our friends and neighbors Claudette and Jonathan Beit-Aharon had relied on Sunlight Solar for the twenty-four-panel PV array they had installed on their house two years earlier, and they were happy with the results. Not only were they able to cover their own power needs, but our local electric utility, NSTAR, allowed the Beit-Aharons to transfer several hundred dollars a year of surplus power as a cash credit to the account of their son Noah, who lives in an apartment a few miles away. Claudette and Jonathan's endorsement counted for a lot as we groped our way toward this brave new solar world.

Even more reassuring were the honest and straightforward answers that Sunlight Solar's sales representative, David Radzihovsky, gave to our many questions. David described in general terms how our roof's orientation (about 50 degrees east of due south) would affect the output of a solar array. We would get about 10 percent less power than an ideally oriented south-facing roof, he said. The main roof's very steep pitch would further reduce output, but only slightly. And shading from the house might knock another few percentage points off the output of a second set of panels on our garage. In all, he said, annual output from a PV array mounted on our house and garage would fall about 15 percent short of optimal productivity.

That was the sobering news. Then David ran through some happier numbers, citing the federal and state incentives that we could tap for a solar installation sized to meet most of our home's power needs. Along with a 30 percent federal tax credit, we would qualify for a $2,000 state rebate on the purchase price of our system. In addition, once our solar panels began generating power, we would accrue market-traded Solar Renewable Energy Certificates (commonly known as SRECs), purchased by retail electricity suppliers to keep pace with their obligations under the state's mandatory phase-in of renewable energy. (The Massachusetts renewable portfolio standard or "RPS" ratchets up the minimum amount of electricity that must come from "new" renewable energy sources, reaching 15 percent of all retail sales by 2020.)[2] In the first year, David told us that we could expect about $1,800 for the SRECs we were likely to earn from our solar generation. All those benefits, of course, would be on top of the money we would save by shrinking our monthly electric bills.

The following day David sent us an itemized quote for a 5.9-kilo-watt solar energy system that would provide about 87 percent of our total electricity needs, averaged over the year and based on our prior year's utility bills. The nominal cost of this twenty-three-panel installation was just shy of $27,000, but we would end up paying about $17,000 after availing ourselves of all the state and federal incentives. As for the payback, David predicted that we would recoup our investment within five to six years.

Comparing notes with Claudette and Jonathan, I was stunned by how much the price of solar had dropped in the two years since they had purchased their PV system. For a rooftop power plant nearly identical in size, their sticker price was over $40,000, and after all the incentives, they shelled out about $25,000. It will take them roughly eight years to earn back what they paid.

Our PV system was the seventeenth deal David Radzihovsky had closed since joining the Sunlight Solar sales force six months earlier. He signed up with the company as his first job after graduating from the University of Massachusetts, where he had reshaped a political science major into a self-styled study of sustainability. His inspiration was a professor who grew most of his own food on a farm near Amherst. Though he respected his professor's dedication to self-sufficiency, David knew that his own outgoing personality would lead him elsewhere. "I was not very much into farming. I was more into the social aspect of sustainability," he told me.

David earned money between college semesters as a Ford and Chevy salesman at two local car dealerships. When he graduated, he wanted to put his sales skills and natural extroversion to work in an area less at odds with his environmental passions. Selling solar systems was a good fit. As we navigated our financing options and technology choices, his crisp explanations were a great help. We could buy the system outright or, to avoid a large cash outlay, he told us we could sign what's called a power purchase agreement, or "PPA" as it's known in the trade.

PPAs and leases have taken the US solar market by storm in recent years, bringing solar power within easy reach for millions of homeowners who aren't able or willing to make an upfront investment that typically begins at around $10,000 and can grow to over $20,000 for larger systems. Under a lease, the homeowner pays a fixed monthly fee but then gets all the power generated by the solar panels at no added charge. With a PPA, the homeowner pays nothing on the front end, but is obligated to buy all the power generated by the solar installation at an agreed-upon price—typically below the retail price charged by the local utility. Under both leases and PPAs, homeowners retain rights to all the power generated by their solar arrays, with any surplus power counting as a credit on their electric bills. Leases and PPAs run for twenty to twenty-five years, corresponding roughly to the length of time solar panels are expected to operate at a relatively high level of productivity. (A typical solar array loses half-a-percent or less of its output per year, leaving most of its generating capacity intact even after a quarter-century of operation.) When the lease or PPA expires, the homeowner has the right to buy the solar system outright, at a price reflecting its vastly reduced value, or to instruct the third-party owner to remove the panels at no cost.

We opted to buy our solar system. The terms of Sunlight Solar's PPA looked good, but we felt more comfortable owning the equipment and found the short payback period attractive. When it came to selecting solar hardware, we chose what David called "the Toyota of solar panels"—a very reputable product made in America by a German-owned company called SolarWorld. Our alternative was the "Mercedes," a pricier, top-performing panel made overseas by the California-based SunPower Corporation.

Once we had signed the contract, David and his team set about applying for a local building permit and filed all the necessary papers to qualify for the Massachusetts rebate. The process was incredibly easy; all we had to do was write a few checks and sign a handful of documents. Within a few weeks, the necessary approvals were in hand.

—⁓—

I remember well the blustery Friday afternoon in mid-March when I peered out of my home-office window at the foot of snow that had fallen since the previous evening. Our roof was blanketed in a thick coat of white. Would the Sunlight Solar crew show up the following Monday, as planned?

After a surprisingly mild weekend, two gleaming white vans pulled into our driveway. The crew emerged: four installers led by a foreman with decades of experience as a home builder, plus a master electrician. Craning their necks, this rugged-looking group of men gazed up at the roof they were about to negotiate. Though clear of snow, its 55-degree slope was no laughing matter. The youngest among them, Liam Madden, admitted that this would be the steepest roof he had ever worked on. His jaw looked taut as he coiled a heavy nylon rope and fitted himself with a climber's safety harness. Having served in a Marine Expeditionary Unit in Iraq, this brawny young guy was accustomed to risky jobs, but today's work would test his nerves in a different way.

Before long, Liam and the other installers were up on the roof, suspended like intrepid mountaineers from ropes anchored to its peak, more than forty feet in the air. In near-silence, they rappelled down the slope and braced themselves as they drilled through the shingles into our roof beams, attaching anchors for the slender aluminum rails that would hold nine panels in place. Then, ever so carefully, a pair

Sunlight Solar Energy installers brave the roof of the author's
home in Newton, Massachusetts. (Photo by author)

of men grasped each poster-sized panel, walking it down the roof's slant and clamping it into its assigned spot. Several passersby stopped their cars to witness this high-wire spectacle. A few even recorded it on their smartphones.

The panels blended beautifully with our roof's charcoal-colored shingles, creating a subtle geometric pattern that ran along the ridgeline, encircled a tall chimney, and sloped down the southeastern section of the roof. Seeing the panels in place allayed my fears that neighbors on our street, with its traditional Victorian homes, might object to the arrival of this modern technology.

The next morning, the Sunlight Solar crew found it a snap to install a second solar array on our garage's more gently angled roof. They bantered with one another and enjoyed music from a boom box while installing fourteen panels in two even rows. In just a few hours, the panels were in place and the wiring was ready for Sunlight's master electrician to hook up all twenty-three panels to the combiner box he had installed in our basement. It took a few weeks for our local building inspector to approve the completed installation, but by early April we were drawing our first electrons from the sun.

—m—

Once we had installed our solar system, I found myself in conversation with several friends and neighbors who had already installed solar arrays or were thinking about doing so. One of those friends was my first post-college boss, Chris Palmer, who hired me to work as an energy policy advisor to Senator Charles Percy, a moderate Republican from Illinois who had previously served as CEO of the Bell & Howell Corporation.

A technology buff who was concerned about America's growing reliance on foreign oil, Chuck Percy was one of the Senate's leading proponents of energy conservation and renewable energy. By the time I joined his staff in the summer of 1978, he had already distinguished himself as a cofounder of the Alliance to Save Energy, together with Minnesota Democrat Hubert Humphrey. (It may be hard to imagine, but in the years following the Arab oil embargo of 1973–74, Republicans and Democrats worked in substantial harmony on many issues, including the adoption of laws to promote more sustainable energy policies.)

Chris Palmer, like Chuck Percy, was fascinated by the advent of solar energy. A former British naval officer, Chris had studied engineering and naval architecture before immigrating to the United States in 1972. On Capitol Hill we worked together to craft legislation that would open up federal support for a number of exciting new energy technologies, including solar and wind, that were emerging from university labs, corporate research divisions, and backyard workshops.

Remembering those heady years, I was delighted when I received an e-mail from Chris, excitedly telling me that he and his wife, Gail Shearer, had recently installed a PV array on their home in Bethesda, Maryland, just outside the nation's capital. "I've worked in the environmental and energy field for 40 years and I've always promoted renewable energy," he wrote. "I'm very happy to be finally doing my part in my home—no longer just talking the talk!"

A few months later I visited Chris and Gail to hear more about their new solar investment. After greeting me on the porch of their colonial-style home, Chris guided me upstairs to a window with a view of the panels neatly spread across their gently sloping, south-facing roof. Gail then joined us at their kitchen table, where we talked about the factors the couple had weighed in deciding on solar. I learned that they had actually entered the solar era a full decade earlier, when they hooked up a few solar water-heating panels to help meet their home's hot water needs. That had been a hasty move, Chris admitted, and the cut-rate panels they purchased had yielded spotty results. Several years later, when the Shearer-Palmers began hearing that the price of solar PV systems was dropping at a breathtaking rate, they were not about to make another impulse purchase. After one false start with an online solar developer, they invited a sales representative from SolarCity, the nation's leading residential solar installer today, to drop by their home for an initial chat.

"We were tough customers," Gail recalled. The SolarCity salesman met with Chris and Gail several times to run through the different options for buying or leasing a system, and, after three years of fence-sitting, the Shearer-Palmers finally signed a contract. "He deserved whatever commission he finally got," Chris admitted. "He earned it."

The arrangement that Chris and Gail settled on has an interesting twist. They could have signed a power purchase agreement with no upfront payment and a fixed monthly charge of $94 for all the power generated by the PV array on their roof. Instead, they opted for a lease with a one-time payment of just under $12,000, supplying them with solar electricity free of any additional charge over the twenty-year term of the agreement.[3] According to the company's calculations, this would be the equivalent of buying power for about 6.6 cents per kilowatt-hour—well below the current rate charged by the Shearer-Palmers' local utility and certain to be an even greater bargain as the years go by and utility rates inevitably rise. Like the Beit-Aharons, Chris and Gail expect to recover their initial outlay within eight years, and by the end of the lease period, their household should have saved close to $30,000 through lower electric bills. When Chris boasted that this was a great "ROI," Gail bantered that her husband, a nonprofit wildlife film producer for most of his career since our shared days on Capitol Hill, has suddenly become "Mr. Return-On-Investment."

"Before solar, our electric bill averaged about $200 each month," Chris told me. During their system's first year of operations, their lowest monthly bill was $12 and no single month's electricity cost them more than $50. Gail added: "We're approaching retirement, and we thought, 'Well, better to pay in now and then face lower electric bills when our income goes down.'" If they end up selling their house before the lease period expires, their buyer will get the benefits of a fully paid lease, which Chris and Gail hope will be reflected in a higher sale price. According to researchers at the Lawrence Berkeley National Laboratory, this hope is likely to be fulfilled. Using a technique called "hedonic" marginal valuation, the Berkeley Lab team looked at thousands of solar-powered homes in multiple locations and found that those with an average-sized PV system earned a price premium of roughly $15,000 over non-solar homes.[4]

Along with liking what solar energy has done to their electric bills, Chris and Gail speak about an added benefit: much greater awareness of the power they're producing and consuming. Gail often goes online to check the output of their solar array, and she finds herself much more bothered than she used to be when her three adult daughters come home for the holidays and leave lights on everywhere. The 71 percent of their power that now comes from solar may be impressive, but she wonders, "Can we do better next year?"

Chris finds himself thinking back to the sacrifices his parents made in England during and after World War II. "Waste was anathema," he recalls in his still-crisp British accent. "When my father took a bath in the Navy, a guy came around with a ruler, and the water couldn't be more than three inches deep. If it was, you were punished." That austere period in Britain, with its victory gardens and fuel rationing, may seem utterly foreign to most Americans today, but Chris translates those family memories into a more contemporary concern about the resources we waste and the environmental damage we cause by generating so much electricity—mostly from fossil fuels. "There's just this desire not to squander something we care about," he says.

—m—

According to the Department of Energy, roughly a quarter of the eighty million owner-occupied homes in America are a good match for solar. The roofs of these buildings have good southern exposure, are not obscured by shade from trees or other buildings, and do not face use restrictions because of shared apartment or condo owner-ship.[5] Despite these solar-friendly attributes, many homeowners can't afford to buy their solar arrays outright, as we and our friends the Beit-Aharons did, or to pay an initial leasing fee of several thousand dollars, like the Shearer-Palmers. To surmount this financial hurdle, many homeowners have turned to third-party power purchase agreements or leases that require little or no up-front payment.[6] Under these arrangements, homeowners have access to solar power at an agreed-upon price for decades to come. The solar companies offering leases and PPAs enjoy their own set of benefits. First, they qualify for the very substantial 30 percent federal investment tax credit that applies to new solar installations, though this credit will only be available for residential solar systems through the end of 2016, unless Congress renews it. They also benefit from the up-front rebates that Massachusetts and many other states offer to solar equipment purchasers. These rebates are also being phased out in a number of states, however, partially in response to an otherwise encouraging development: the steady decline in the cost of solar panels and installed solar systems.

By the end of 2014, solar arrays had reached nearly 600,000 American residential rooftops—about one in every two hundred households.[7] That includes the millions of multifamily residences and rental properties that solar power has barely touched. The number of solar households is expanding rapidly, and much of that growth is in solar arrays that are not owned by the households where they are installed. In California, which leads the nation in residential solar installations, more than two-thirds of all new solar contracts are for third party–owned systems—up from only 10 percent in 2007. These installations alone gave nearly a billion-dollar boost to the California economy between 2007 and 2012.[8] In Arizona, Colorado, and New Jersey, new third party–owned systems have hit or exceeded 90 percent of new installations in some recent periods, and in my home state, about half of all new solar systems are third party–owned.[9]

Leases and power purchase agreements have extended solar energy's reach to many middle-class families, but what about lower-income households? Too often, these families are left out in the cold because they fail to meet the thresholds for credit-worthiness set by solar companies and their lenders. Recognizing this gap back in 2006, California mandated that over $200 million be spent over a ten-year period installing free or heavily subsidized solar arrays on lower-income homes. In 2013 another $108 million was added to the program. California's three big investor-owned utilities are required to raise these funds through a surcharge on their customers' electric bills.[10]

An Oakland-based nonprofit called GRID Alternatives has taken the lead in implementing California's Single-Family Affordable Solar

Housing program (SASH), delivering solar energy to about 1,200 low-income, single-family households per year. To see how this group goes about its work, I spent a morning with one of its installation crews in San Francisco's Bayview district, near Candlestick Park. I was met outside my downtown hotel by Cathleen Monahan, a North Carolina–born glass artist-turned-community organizer who directs the SASH program. As we drove in a borrowed, beat-up red pickup to the work site, Cathleen introduced me to her organization's impressive work.

America's biggest nonprofit solar installer, GRID Alternatives, uses a small paid staff to leverage an impressive amount of on-site training and volunteer commitment. Cathleen describes her organization as a "teaching hospital for solar," estimating that eleven thousand people have learned the basics of solar installation since the group began implementing SASH in 2009. Some of those trainees have since moved into paid positions at for-profit solar companies, but many have stayed on as volunteers. "A lot of folks just come because they enjoy giving back to the community," she tells me.

Originally GRID Alternatives operated solely out of its Oakland office, but to broaden SASH's reach it has opened up six additional offices around the state. Recently, with a $2 million grant from Wells Fargo, it expanded its operations to Colorado, and in 2014 it began gearing up for work in the New York–New Jersey area, as well as in Washington, DC.

As Cathleen guided us through highway interchanges and crowded city streets, I did my best to concentrate on what she was saying, trying not to be too unnerved by her long glances at the smartphone GPS she held in her hand. Finally we climbed a dizzyingly steep cul-de-sac and spotted a GRID Alternatives truck loaded with solar panels, its front tires braced with wooden blocks to keep it from careening downhill. The truck was parked alongside a small house with an iron-grated door. On its front steps stood the homeowner, Liying Huang, smiling broadly in blue jeans and a flower-printed blouse. I later learned, via translator, that Liying and her husband immigrated to America ten years ago from Guangzhou, China. She is unemployed and her husband earns a modest living as a certified electrician.

The Huangs' house, I noticed, was very small—barely two rooms deep—but it had a flat roof well suited to solar. A wholesome-looking group of volunteers had gathered on the sidewalk, neophytes at construction but ready for a good day's work. The temperature, climbing into the high 80s, was unusually hot for an April morning in the Bay Area, but I learned that San Francisco has microclimates, and Bayview, with its eastern orientation, is on the sunny side of this often fogbound city.

Mara Meaney-Ervin, bubbling with enthusiasm, welcomed the team of volunteers, now standing in a tight circle on the pavement. "It's Earth Week, and installing solar is a great way to celebrate our earth!" she beamed. The solar array to be installed that day will save the Huang family about $16,000 in electric bills over the lifetime of the system, she tells the group. It also will spare the atmosphere about 50 tons of greenhouse gas emissions. "That's the equivalent of taking nine cars off the road for a year, or planting about 1,170 trees," she proclaimed.

As a development officer for GRID Alternatives, part of Mara's job is to seek out Bay Area corporations that are interested in offering their employees a meaningful volunteer experience. This particular work crew—nine people in all—is from Salesforce, a leader in cloud-based computing. Companies like Salesforce are asked to contribute financially to help defray the cost of training their volunteers and installing the solar array. A typical installation's cost may be in the $12,000 to $15,000 range; participating companies generally contribute somewhat less.

Ron Griffin steps up after Mara's introduction. Muscular, with an Oakland Raiders tattoo on one arm and an embellished cross on the other, he's spent the past four years as a site supervisor with GRID Alternatives. Above all, he wants to make sure that the Salesforce volunteers, now wearing nametags on their hard hats, stay safe. Turning to one of them, he asks: "One thing that you do *not* want to do on a roof is what, Jeff?" "Walk backwards," the volunteer responds. Ron credits the correct answer, drilled into the group during an earlier safety orientation back at the Salesforce office. He then lets people know when a hard hat needn't be worn: on the roof. "Gravity says that things are gonna fall *off* the roof and not *on* the roof," he says, "unless you catch a plane comin' by with an old bolt loose or a bird that ate a bad worm. That could get pretty ugly, but hopefully it won't happen."

Ron's levity sets an upbeat tone for the crew as they divide up, some carrying aluminum-framed solar panels into the backyard, others climbing a ladder up to the roof, where they clip their nylon harnesses to a safety anchor. Slowly and carefully, the installation proceeds under Ron's ever-watchful eye. None of the volunteers goes near live electricity; that is reserved for Ron, an electrician who was retrained for solar installation work by an Obama administration–supported "Jobs Now" program. During his safety talk, Ron points to the house's main electric panel and says: "If you see me starin' off into space with my hand inside there, I ain't thinkin' about a beer and I ain't thinkin' about the baby." To dislodge him, he suggests a quick kick or a body block.

Sitting in her compact, sparsely furnished living room, Liying Huang tells me that her home has little need for air-conditioning in San Francisco's generally mild climate. Today is a bit of an exception. As the house is primarily heated with gas rather than electricity, she expects that the new solar array, though small at 1.7 kilowatts, will supply at least 90 percent of her family's power needs. That will yield hefty savings on her electric bills, she says.

Noon approaches, and Liying emerges from her home, shy but visibly delighted with the work in progress. She arranges a picnic lunch on her patio featuring several Chinese vegetarian dishes in aluminum pans—her contribution to the day's effort. Sweaty T-shirted volunteers swirl eagerly around the makeshift buffet before wrapping up their work in the early afternoon.[11]

—⚉—

Programs like California's SASH are beginning to reach across economic boundaries that, until recently, have reserved solar power primarily for middle- and upper-income households. Yet the millions of Americans who are tenants rather than homeowners face different constraints. Roughly a third of all Americans live in rented homes and apartments where they have no legal right to install anything at all on the roofs of their buildings. To make matters worse, landlords have little incentive to generate solar power on their properties—unless, like Brent Haverkamp, they are paying for their tenants' electricity.

Haverkamp owns a four-hundred-unit housing complex adjacent to Kirkwood Community College in Cedar Rapids, Iowa. Several years ago, he decided that the best way to market these rental apartments to students was to offer a "hassle-free living" package, with all utilities folded into the monthly rent. That left him with a hefty electric bill of $275,000 per year. By installing two thousand solar panels on his buildings, he has reduced that outlay by a third. Along with the savings on his electric bill, he shrank the system's up-front cost from $1.8 million to under $1 million by claiming federal and state tax credits as well as rebates offered by his local utility. Did idealism play a role in this savvy businessman's decision? "Not at all," he tells me. "I was thrilled to do a clean energy project, but at the end of the day it had to make sense financially for a project of this scale."[12]

Haverkamp's ingenuity could be a trendsetter for solar installations on many larger rental properties where building owners pay for utilities. Vent stacks, skylights, and air-handling units may limit roof space on certain buildings, but parking canopies and ground-mounted arrays may offer expanded solar opportunities at those same sites. However, where none of those options is available, there is another way.

Tenants who have no roof access, homeowners whose roofs get inadequate sun, people with aging roofs that need to be replaced before a twenty-five-year solar array can be installed—all of these are prime candidates for an innovative alternative variously called "shared solar" or "community solar." Community solar programs allow renters and owners of single- and multifamily homes as well as businesses and nonprofits to buy shares in an off-site, shared solar facility rather than installing solar panels on their own premises. Also referred to as "solar gardens," these facilities in some ways resemble community-supported agriculture, an increasingly popular way for people to buy shares in the output of a locally run farm without actually owning the farm equipment or the land on which their food is grown. But instead of yielding fresh seasonal produce, solar gardens produce power that is credited to shareholders' utility bills in amounts proportionate to their investment in the enterprise.

Shared solar arrays have a number of advantages relative to PV panels placed on individual homes and other small buildings. Tom Sweeney, chief operating officer of the Colorado-based Clean Energy Collective, outlined a few of them when he spoke at the solar industry's annual trade conference, held in Chicago's giant McCormick Place in October 2013. First, he said, it opens up the solar market to a much broader population, as customers can buy very small increments of power rather than having to install a full PV array. Second, "hard" costs for equipment as well as "soft" costs for installation work, permitting, and maintenance are likely to be lower for shared solar systems than for the same amount of generating capacity installed on numerous separate properties. Third, choosing a single site with optimal conditions—southern orientation and no shading, for example—will end up yielding higher output and better financial returns than PV arrays on buildings that were not designed for solar power.[13]

Sweeney's Clean Energy Collective has already built or is now planning more than two dozen shared solar installations in five states. They range from a compact array in Rockford, Minnesota, with only six times the output of our home solar system in Massachusetts, up to a 4,800-panel installation in Paradox Valley, Colorado, rated at more than two hundred times our home's solar output. Any local utility customer can buy one or more panels at these solar farms, and the Collective arranges for the utility to credit the customer's monthly bill with the equivalent share of the farm's total power generation.

To jump-start shared solar, a number of states have passed laws requiring utilities to offer this particular form of green energy to their customers. Under a California law adopted in 2013, the right to buy into shared solar programs is extended to homeowners and renters, as well as local governments and businesses. Fully implemented, this Green Tariff Shared Renewables Program could supply up to 600 megawatts of power to the grid through community-scale solar farms—enough to meet the electricity needs of 130,000 California households.[14] And of this total, at least 100 megawatts is specifically earmarked for low-income residential customers.[15]

—⁓—

Back at our home just outside Boston, spring has finally arrived. During a winter of prolonged, bitter cold and relentless snows, there were full weeks when the solar array on our garage roof remained blanketed by snow. On our house, the roof's steeper slope caused the panels to shed snow more quickly, but there too, days would pass with no sun filtering through. I didn't have to venture outdoors to make these observations; all I had to do was check my online monitoring program to see how little solar power we were producing.

Despite the rough winter, our two solar arrays have performed admirably overall. During the past year, we have drawn about 75 percent of our power from the sun. This was lower than the 87 percent that Sunlight Solar's David Radzihovsky predicted when he first surveyed our roof and looked at our previous year's electric bills. One cause is the snow. The other is our car. Just before our solar panels went up, we bought a blazing-red Ford C-Max Energi, a plug-in hybrid electric vehicle that now meets all the commuting needs of my wife, Tamar, and most of our family's shopping and leisure requirements. We fill the C-Max's gas tank about four times a year and it's averaging more than 150 miles per gallon.[16]

We love the electric car, even though it has created a minor, ecologically induced strain on our marriage. When Tamar comes home from work, she invariably plugs the car in for a recharge, even if she's going out again in the evening. I try to tell her that she only gets one charge per day, as I want to be able to boast about how much of our electricity we're getting from the sun. Those extra electric miles are driving down my solar score!

While I generally lose the battle over multiple car charges, Tamar is fully on board with declaring our electric clothes dryer out-of-bounds. An outdoor clothesline in the summer and indoor racks in the winter have eliminated this appliance from our household energy budget over the past year. We've all but stopped using central air-conditioning. And like Chris and Gail's visiting daughters, our own two daughters, Tali and Maya, know that leaving lights on in their empty bedrooms, or downstairs after hours, is taboo.

These efforts at conservation have almost, but not quite, canceled out the added power demand of our Ford C-Max. Daily car charges account for a quarter of our electricity consumption, but our total power needs have risen by a much more modest 3 percent. Clearly, heavy snows have been the real culprit in reducing the sun's share of the electricity we use. Yet even with all the bad weather, the 5,790 kilowatt-hours of solar power we generated over the past year have spared the globe 3.6 tons of CO_2 emissions and have offset the equivalent of 520 gallons of gasoline or 1,350 electric washer-and-dryer loads. Thinking in terms my college-age daughters might appreciate, we've generated enough solar electricity to charge over a million smartphones![17]

Our financial savings have also been considerable. Given the 16 cents per kilowatt-hour that our local utility has charged us over the past year, we have shaved about $950 off our electric bills. On top of that, we have earned about $1,400 in Solar Renewable Energy Certificates (SRECs). For every megawatt-hour (1,000 kilowatt-hours) of solar power we produce, we earn one of these certificates. Our state's electricity suppliers must buy a sufficient number of these SRECs to meet their assigned obligations under the Massachusetts renewable portfolio standard.

At first I was disappointed to learn that our Massachusetts SRECs have yielded lower returns than Sunlight Solar had factored into our yearly payback projection.[18] I console myself by remembering that there's an upside to this development: the lower value of these certificates reflects a very robust statewide investment in solar energy, which in turn has made it less expensive for utilities to buy the SRECs they need to meet their renewable energy obligations. If electric rates and the traded price of SRECs remain constant, we'll recoup our solar investment in a little over seven years.

There's another factor, though, that affects the timing of our investment return: the price of electricity. New England's biggest utilities have already been granted approval to raise their electric rates—largely the result of a surge in demand for natural gas that is taxing the capacity of existing pipelines to deliver sufficient amounts of this fuel to the region.[19] If new pipelines are built, their cost will be passed on to consumers; if greater amounts of natural gas are brought to New England on LNG tankers, that too will drive up the price of gas delivered to our power plants. This may strain household energy budgets, but it will also increase the value of home-generated power in the years ahead. Taking this into account, we may recover our up-front solar system costs in the five to six years that David Radzihovsky predicted after all.

—⚏—

A conversation with Paul Israel, president of Sunlight Solar Energy, reminds me of how far solar power has come over the past few decades. In the mid-1980s, Paul left a journalism job in Washington and headed west. For a while he sold and installed residential solar water-heating systems in California, taking advantage of the generous rebates offered by the Sacramento Municipal Utility District (SMUD). Then he found himself in Quartzsite, Arizona, a hamlet in the Sonoran Desert where thousands of snowbirds from the Pacific Northwest and Canada would spend winters in their recreational vehicles. "After a while they were sick and tired of slogging fuel to keep the generator running, and they'd ask me, 'Hey, how can I cut down on my generator use?' Solar was the answer." He began installing PV panels on these motor coaches—just big enough to recharge a 6- or 12-volt battery. Gradually he built a $10-million-a-year business selling residential solar systems. So how does he assess his own company's progress, and that of the solar industry in general? "I'd call us teenagers, but we're definitely moving forward into young adulthood," he says.[20]

No longer an expensive novelty, solar power is quickly gaining acceptance as a welcome substitute for big utility bills and a way for homeowners like us to do something *personally* to trim America's overdependence on carbon-based fuels. But homeowners aren't alone in embracing solar energy. On warehouse roofs, mall parking lots, sports stadiums, university campuses, and government buildings, new solar installations are slashing electric bills and are proving that central-station power plants fueled by fossil and nuclear fuels aren't the only avenues to prosperity. Exploring some of these new installations would be the next stop in my solar journey.

CHAPTER TWO

Ballfields and Boxtops

TO MOST NEW ENGLANDERS, Gillette Stadium is home to the beloved Patriots. On a typical home-game weekend, nearly seventy thousand of them pile into their SUVs, pickups, and minivans and make the traffic-snarled pilgrimage to Foxborough, 30 miles south of Boston. There, in sprawling parking lots, they fire up their grills, open up some beers, and toss a few footballs while they wait for game time.

My first Gillette experience was a bit different. On a balmy mid-summer night, I went to a Taylor Swift concert with my daughter Tali. I'm sure I was the only attendee whose excitement about the concert was eclipsed by fascination with the sleek solar canopies that shade the walkways at Patriot Place, a retail mall adjacent to the stadium. As Tali and I wandered through this outdoor mall, translucent matrices of solar cells hovered overhead, delicately clasped to airy steel trusses that spanned the broad, landscaped passageways. These largely ornamental cells, together with more conventional panels atop surrounding buildings, generate about 20 percent of the shopping complex's year-round power needs.

Solar power is growing in popularity at sports arenas across the country, spurred on by one of the nation's most forward-thinking power companies, NRG Energy. Just before the Super Bowl in February 2011, the company's CEO, David Crane, gathered a group of NFL owners in a Dallas hotel and invited them to join him in creating icons for the new solar era. In true competitive spirit, eight teams signed on almost immediately. By mid-2014, five NFL stadiums had solar power, with a sixth array slated for completion by the end of 2015.[1]

At the invitation of Tom Gros, chief customer officer at NRG Energy, I returned to Gillette to watch the Patriots play the Dolphins on a chilly Sunday in October. After the game, Tom guided me through his company's business strategy as we descended from NRG's corporate suite and walked out onto the artificial turf. Football is the sport that senior executives at Fortune 500 companies most frequently attend, he said, and what better way to reach them than through the owners of NFL teams? "The NFL is a collection of family-owned firms," he explained, "and those businesspeople are gatekeepers in their communities."

Tom is openly enthusiastic about solar energy's promise of producing low-carbon electricity. An aerospace engineer by training, he enjoys puzzling over how to adapt solar PV technology to sports arenas and other commercial settings. The Patriot Place arrays are relatively simple; most of them sit on conventional flat-roofed buildings several hundred feet from the stadium itself. Mounting solar panels directly onto stadiums demands greater technical agility. For Tom, this hit home when he attended his first Eagles game at Philadelphia's Lincoln Financial Field. Every time the Eagles scored, a round of fireworks burst into the air from a series of ballistics boxes placed along the top rim of the stadium. Tom's boss turned to him after the first set of explosions and asked: "Why are you so pale?" Tom responded: "Because that's where we had planned on putting a large solar array!"

The NRG design crew devised a clever way to protect the solar panels' tempered glass surfaces: both the ballistics boxes and the panels themselves were equipped with shock-absorbing mounts. But Tom was still concerned that sudden flashes of light from the fireworks might damage the highly photosensitive solar cells embedded in the arena's eleven thousand panels. He persuaded the solar manufacturer to certify that this would not cause a warranty issue for the panels. "No other solar developer in the world has ever had to think about those elements," he told me as we strode through the thinning crowds at Patriot Place.

At the Washington Redskins' FedEx Field, solar-canopied parking spaces—864 of them—posed a different kind of puzzle. Tom knew that tossing a football in stadium parking lots is a favored pregame activity, and he sensed that having footballs constantly crashing into solar panels wouldn't be a great idea, so he calculated the peak of a typical football's trajectory—13 to 14 feet—and then made sure the solar panels were mounted a few feet higher. Today those shaded parking areas are the stadium's premier parking spots.

The NFL contracts are power purchase agreements whereby NRG retains ownership of the solar installations and sells all the solar electricity at a fixed price to the stadium owners. Tom won't discuss the dollars and cents, but he makes it clear that maximizing output and minimizing cost are not his company's primary objectives. "These are not just power plants," he explains. "They're marketing elements . . . important to the team, to the community, and to my firm as well."

Solar energy is a high-profile feather in NRG's cap, even if it is still a small part of an energy portfolio long on coal, oil, and gas. (NRG is the biggest independent supplier of power to America's electric utilities.)[2] Every year six million people walk through Patriot Place, and NRG's information kiosks are there to ensure that as many of those visitors as possible take note of the elegant solar panels overhead. At FedEx Field, Tom claims that "Solar Man," a 30-foot-high statue lit up by thin-film solar cells, is the most photographed object in the stadium. NRG's goal is to pique people's curiosity, he says. "Little kids can pull their dad aside and say, 'Tell me what's going on here.'"[3]

Stadium solar may be long on symbolism, but many of America's biggest corporations are getting serious about shifting to solar and other renewable energy technologies for their power needs. Apple plans to use renewable resources to meet 100 percent of its energy needs at the "Apple Campus 2" in Cupertino, California. About 11 percent of the power for this research and office complex will come from on-site solar photovoltaic arrays; about half will come from on-site fuel cells powered by biogas; and the rest will be purchased from off-site renewable energy projects, including utility-scale solar farms.[4] Google has set its sights on someday powering all of its operations with renewable energy. Already the solar panels on rooftops and parking structures at its headquarters campus in Mountain View, California, meet 30 percent of the complex's peak electricity demand.[5]

It's predictable that high-tech industry leaders would be turning to solar energy. More surprising is the investment that traditional American companies like Costco, Kohl's, Macy's, Staples, Toys R Us, Walgreens, and Walmart are making in solar. In recent years, these and other mega-retailers have begun blanketing the rooftops of their stores and warehouses with solar panels. Walmart is in the lead so far, with rooftop arrays on more than 250 of its stores, adding up to more than 100 megawatts of power-generating capacity.[6] The company is committed to bringing solar energy to a thousand of its properties by 2020, and in the long run it has declared that 100 percent of its company-wide energy needs will come from renewable sources.[7] Walmart is giving a high profile to its solar investments as a way to repair its often-battered public image. Ironically, though, the Walton family—Walmart's majority owners—are jeopardizing this newly garnered goodwill through their contributions to organizations that are leading the charge against pro-solar policies in a number of states.[8]

Though less in the news, thousands of smaller businesses, schools, and nonprofits across America are capping their rooftops and parking lots with solar panels. I traveled to Lawrence, Massachusetts, once known as the shoe capital of the world, to see one of those installations. In this struggling city, hulking red brick mill buildings line the Merrimack River, many of them abandoned or minimally occupied. A network of canals runs through the city, originally supplying hydropower to the factories that fueled the city's growth a century or more ago. One of these narrow waterways flows alongside a handsome two-story building at 468 Canal Street, once occupied by a firm that manufactured worsted wool. Today the building is owned by Cardinal Shoe, a family-owned business with deep roots in the Lawrence business community.

Cardinal Shoe's president, Richard Bass, is proud that his family's company has withstood the industrial exodus from Lawrence. At one point the company employed 375 people and produced 300,000 pairs of women's dress shoes per year. That was until the lure of cheap labor drove most of America's shoe manufacturing overseas. Wanting to stay local, Cardinal adapted by downsizing and developing a successful niche: the production of custom-ordered ballet slippers for many of the world's leading dancers.

Richard is dressed well for the early summer heat, in a white short-sleeved shirt and matching white trousers. His immaculately coiffed, jet-black hair belies his seventy years. He guides me to the factory floor, where fifty workers are busy molding, sewing, and gluing hard-toed *pointe* ballet shoes. After showing me one sewing machine operator's fine stitchwork, he introduces me to his son and daughter, who have recently joined the business. Skilled in computer-aided design, they now spend much of their day in front of computer screens, processing orders from clients as far away as Moscow. In another room, enlarged photos of famous dancers line the walls. Richard points to one and says, "This is probably the best girl dancer in Russia. She used to wear Russian shoes, and she came over to wearing ours."

Richard is also proud of the solar array that he decided to install on the roof of his building. "We are the largest solar-powered ballet shoe factory in the world," he tells me. Then he pauses. "Of course, we are the *only* one!" With youthful agility, he leaps up the rungs of a tall ladder that leads us onto the roof, where we get a clear view of the mill buildings downriver and across the canal. "These buildings used to get their power from below," he recalls. "We had the river here creating the power." As we walk to the western half of this 800-foot roof, our conversation traverses a century-and-a-half of energy technology, leapfrogging over the fossil fuel era. "Here we are at the other end of the spectrum," Richard says, pointing to long symmetrical rows of solar panels—over a thousand of them mounted on thin metal rails with an ever-so-slight southward tilt.

Because of the roof's rubber surface, capping several inches of foam insulation, no holes were drilled when the solar array was installed. Instead, the panels are weighted down, or ballasted, with evenly spaced cinder blocks. On many flat-roofed buildings I've visited, this is the preferred option, securing solar assemblies against high winds while keeping waterproof roof membranes intact. In some cases, roof beams need to be reinforced to carry the extra weight.

The solar array on this old mill building not only supplies all of

Cardinal Shoe's electricity, but it also meets the power demands of Richard's twenty-seven tenants—a variety of smaller businesses that occupy the lower floor. Better still, Richard paid nothing up-front for his solar installation. Instead, he signed a power purchase agreement with Revolution Energy, the company that owns the array, guaranteeing him all the electricity it produces at 7.5 cents per kilowatt-hour, little more than half the rate he was paying for power from the grid. Throughout the twenty-five-year term of the contract, that price can rise by no more than one-tenth of a percent per year.[9]

—ɷ—

Kathleen Doyle, CEO of FireFlower Alternative Energy, was a real estate broker for twenty years, building what she describes as the largest woman-owned commercial brokerage in Massachusetts. This was a big leap for someone who grew up in a public housing project in Boston's Brighton neighborhood. Today, as the developer of solar power systems for commercial and industrial buildings, she deals with many of the same clients. "It's still an old boys' network," she tells me, but she clearly enjoys being a pioneer in opening up this circle to a little gender diversity. In her spare time, she also likes mentoring army veterans and other young people who are trying to launch their own business careers.

I met Kathy at the Boston-Dedham Commerce Park, where she recently put up a solar array about three times the size of the Cardinal Shoe installation. Used as a train repair facility when it opened a century ago, this 450,000-square-foot building now includes a commercial bakery, a printing company, a few small manufacturers, art studios, and warehouse space. By relocating a few heating and air-conditioning units, Kathy made room for 3,300 PV panels on the building's linear roof, supplying about 65 percent of the building's power load. From the air, the installation looks like an elongated electronic circuit board.

Kathy's savvy in putting together real estate deals has helped her succeed as a solar developer. Working on a commission paid by building owners, she gets the necessary permits, lines up equipment suppliers and installers, locks in the state and federal incentives, and then delivers a completed project to her clients. While she can't reveal the specifics of her contract with First Highland Management & Development, the Commerce Park's owners, Kathy tells me that she typically reserves the option to become a part owner of projects she develops—up to 15 percent of total equity.

When Kathy first approached executives at First Highland, they were hesitant to deal with the headaches of owning and managing this new technology. "I'll do all of the work," she promised them, "and you don't have to pay me until you're producing kilowatt-hours." One of the first things she did was to line up a federal grant covering 30 percent of the installation's nominal cost of $4.2 million. Grants like this were authorized following the economic crash of 2008, on the assumption that solar and other renewable energy investors during that difficult period might not have sufficient income to take advantage of the tax credits made available under other federal programs. Known as Section 1603 grants (referring to the relevant section of the American Recovery and Reinvestment Act), these federal subsidies gave a major boost to the US solar industry during a critical stage in its commercialization.[10]

Just before Christmas in 2012, Kathy handed a completed 974-kilowatt project to First Highland, and she was thrilled that her teenaged son agreed to be on hand to help with the ribbon-cutting ceremony. "For an eighteen-year-old to be impressed with his mother is a huge coup," she recalls of that day. Given her solar projects and a sideline business she shares with her husband making biofuel from waste vegetable oil collected at area restaurants, Kathy's guess is that their boy, now in college, will end up in the energy business.

Now that the Commerce Park's solar array is up and running, its owners benefit from all the free power it generates. If, in any given month, the panels produce more electricity than the building consumes, the net-metering provisions under Massachusetts law allow those extra kilowatt-hours to be carried over to the next month. If there is a surplus at year's end, the local electric company is obligated to pay the owners for that power at the full retail rate. With a system significantly smaller than the building's current annualized load, it's not likely that the Commerce Park's installation will produce a year-end surplus. However, if several of the property's bigger tenants were to leave the building at any point, net metering would ensure that any excess solar power would be well compensated.[11]

SRECs offer an additional benefit to solar power investors like First Highland. Electricity produced by the rooftop array at the Boston-Dedham Commerce Park earns SRECs that are sold to the state's electricity suppliers, at a value just slightly lower than the rate per megawatt-hour assigned to home-mounted solar arrays like the one on our house in Newton.[12] With solar hardware and installation costs dropping, New England electric rates rising, and a revised Massachusetts SREC program that promises to bring greater long-term stability to the market for solar certificates, energy entrepreneurs and building owners alike have good reason to embrace solar energy as a way to reduce the carbon footprint of commercial properties in the Bay State.

—⁂—

New Jersey is widely regarded as the birthplace of modern solar power, dating back to the much-heralded production of the first silicon solar cells at Bell Laboratories in the 1950s.[13] Somewhat improbably, this little mid-Atlantic state has regained its solar preeminence in recent years. With the highest population density of all our states, the fourth-smallest land area, and much less solar radiation than most of the West and much of the South, you would think New Jersey would be at or near the bottom of the heap in installed solar power.[14] Yet it ranks third in the nation in total installed solar PV capacity, trailing California and Arizona but ahead of all other Sun Belt states.[15]

New Jersey's strong pro-solar policies, primarily driven by a successful but stormy SREC program, are responsible for the state's remarkable solar track record. The residential solar market has flourished, the state's biggest utility has mounted solar panels on 200,000 of its power poles (an application less than affectionately known as "solar on a stick"), and a growing number of commercial and industrial property owners have come to recognize the solar value hidden atop their buildings and warehouses.

Avi Avidan was born and raised in Israel, where solar water heaters were a pervasive presence on the rooftops of homes and apartment buildings long before their use was mandated by law in 1980. Fuel shortages, the high price of conventional power, and Israel's sunny climate made these simple units an obvious choice. After immigrating to America in 1974, it was a long time before he returned to his solar roots. By the time he did so, he had built up a substantial real estate business that he now manages in partnership with his American-born son Josh.

The Avidans own or manage eleven industrial spaces totaling roughly three million square feet. In 2011, they installed a 17-acre solar power plant on one of their warehouses in Edison, New Jersey. At the time, Avi says, it was the largest rooftop solar power plant in the country, and probably the world.

From outside, the warehouse at 145 Talmadge Road looks relentlessly bland. Truck bays line three sides of the building, and an imposing, windowless brick façade greets visitors as they pull into the parking lot. Avi and Josh meet me there and lead me to a small interior office with a window air conditioner rattling directly overhead. They speak with blunt pragmatism about the experience they've had installing solar energy on two of their properties.

Facing a soft real estate market in 2010, the Avidans began brainstorming about ways to attract new tenants to the Talmadge Road property. "We had to do something to make our building stand out and be something different than most industrial buildings out there," Avi said. Solar struck him as a good bet, but only if it were accompanied by major steps to improve the building's energy efficiency. Spending about $5 million, they insulated walls and loading dock doors, found ways to capture and reuse waste heat, installed smart climate controls, invested in energy-efficient lighting, and replaced the building's giant asphalt roof with a foam-insulated rubber roof like the one

Josh Avidan (left) and his father, Avi, developers of a 4.26-megawatt solar roof
atop a food storage warehouse in Edison, New Jersey. (Photo by author)

at Cardinal Shoe. Altogether, Avi estimates that these measures cut
the building's energy consumption by about 30 percent.

Then came the much bigger expense of capturing enough solar
energy to meet about half the building's remaining power needs, including a large frozen storage area for tenants who are in the food
distribution business. The Avidans blanketed the building's roof with
17,745 PV panels capable of generating over 5 million kilowatt-hours
of electricity per year. The price tag for this 4.26-megawatt system
was $20 million, reduced by $6 million thanks to a Section 1603 grant
from the federal government. To further lighten the financial burden
of this investment, the Avidans were counting on the sale of roughly
five thousand SRECs per year. When the installation was completed
in April 2011, New Jersey SRECs were selling at $640 apiece. Had the
market remained stable, the Avidans stood to gain over $3 million per
year in SREC payments—far more than the $600,000 they expected
to save annually through reduced utility bills. But the solar market
was glutted with eager investors, and SREC prices plummeted just
months after the Talmadge Road array began operating.

When auction prices for SRECs dropped below $70 in 2012, all who were involved in solar energy development in the Garden State became alarmed. To avoid an industry-wide crash, the New Jersey legislature passed a new law requiring that solar energy supply 4.1 percent of the state's overall power needs by 2028.[16] That helped SREC prices regain some of their value, but not enough to restore the Avidans' initial hopes that they would recover their solar investment in just seven years. Now Avi says: "I'm gonna be lucky if I see my money back in fifteen years."[17]

All New Jersey solar investors have been unsettled by the SREC market's wild swings, but Alan Epstein, president of KDC Solar, is one of many entrepreneurs who continue to view the Garden State's solar market as a promising one. Some of KDC Solar's projects are ground-mounted arrays installed in open fields, but most are giant rooftop systems akin to the Avidans' installation at Talmadge Road. I visited one of these, on the White Rose food storage warehouse in Carteret, New Jersey. The vista from the rooftop of this 800,000-square-foot building is one of hope and despair: hope for a more peaceful, environmentally sustainable America, reflected in the solar panels that stretch the full 1,500-foot length of the building, and despair captured by the view across a gritty tidal strait of Staten Island's Fresh Kills landfill, where all the debris from the Twin Towers tragedy was sorted and is now buried. Epstein tells me that the White Rose warehouse is zero net energy. To meet this standard, its 4.9-megawatt solar array has to generate enough surplus power when the sun is shining to make up for all the conventional electricity consumed at night or when the weather is bad.

As he forges ahead with new projects, Epstein sees the momentous decline in solar panel prices as a sufficient counterweight to the volatility in SREC prices. "We used to buy panels at $2 a watt," he tells me. Now they're little more than 70 cents per watt.[18] "If they were still $2 a watt, we wouldn't be building any more projects."[19]

Opinions vary widely as to how much of New Jersey's power can be drawn from the sun, but I spoke with several key players who agree that the state's solar standard of 4.1 percent by 2028 falls far short of that potential. Assemblyman Upendra Chivukula, an Indian-born electrical engineer and cosponsor of that standard, thinks that 10 to 15 percent of the state's electricity could come from solar, relying principally on rooftops large and small as well as contaminated industrial lands, or "brownfields."[20] Far less restrained is Lyle Rawlings, a solar industry veteran who runs his own solar installation firm and serves as vice president of the Mid-Atlantic Solar Energy Industries Association. Lyle has spearheaded a coalition called NJ FREE (New Jersey for Renewable Energy and Efficiency) whose goal is to rally support for legislation that would require 80 percent of the Garden State's electricity to come from renewable energy by 2050. Of that 80 percent, half would be drawn from solar arrays. In tandem, there would be a mandatory 20 percent reduction in the state's total electricity use.

Lyle acknowledges that NJ FREE's solar agenda would take up a lot of space: roughly 52,700 acres, equal to a nine-mile-by-nine-mile square if all the solar panels were to go in one spot. That's about 1.1 percent of the state's total land area. With efficient use of rooftops, paved parking areas, and other impervious surfaces, he claims it would be possible to reach the coalition's solar penetration goal without encroaching in any significant way on the state's remaining open spaces.[21]

NJ FREE's vision is undeniably ambitious, but it does offer valuable perspective on the resources that would have to be deployed if the Garden State were to move substantially beyond the modest solar goals set under current law.

—⁂—

Giant industrial warehouses like White Rose and Talmadge Road are prime venues for solar development, but many of our college and university campuses offer an even richer rooftop solar resource. It's the rare university campus today that hasn't made at least a symbolic nod in the direction of solar energy, putting a solar array on a building or two. But no campus comes close to the all-out solar commitment made by Arizona State University.

For the 60,000 students who are enrolled at ASU's Tempe campus, just a few miles east of Phoenix, solar energy is part of daily life. Those arriving by car enjoy the shade of solar canopies at more than a dozen parking structures. When students emerge from breakfast or lunch at Memorial Union, a solar parasol soars overhead, shielding them from the sun as they gather with friends in a landscaped courtyard. If they study at Hayden Library or head to classes in the Goldwater Science and Engineering Center, they may not be able see the rooftop panels, but a brushed-steel plaque at each building's entrance lets them know that solar power is at work. These ubiquitous signs describe the size of each solar installation, how many kilowatt-hours are being generated annually, and the amount of avoided carbon dioxide emissions, using a metric that every student on this car-commuting campus can easily fathom: the equivalent number of passenger vehicles being taken off the road for a year.

Jean Humphries is responsible for overseeing the construction and operation of ASU's solar installations—all 86 of them. We meet in her office at the University Services Building, on the edge of the Tempe campus. Trained as a librarian, she has created a website that thoroughly documents each and every solar installation on the university's five campuses, right down to the number of PV panels (nearly 96,000 in all), the companies that manufactured them (12 different firms), and the total project cost of each system (ranging from \$138,000 to more than \$11 million).[22] I also learn that NRG Energy—the company responsible for the New England Patriots' embrace of solar power—is the leading solar developer at ASU. Thirty-one of the university's installations are owned by NRG, which sells the solar electricity to the university via a number of long-term power purchase agreements.[23]

Though Jean makes sure that ASU's solar power plants are effectively deployed, she credits President Michael Crow with inspiring ASU's solar energy achievements as part of a university-wide quest for sustainability. In March 2006 he was one of twelve founding signatories of the American College and University Presidents' Climate Commitment, which binds its signatories to "set a target date and interim milestones for becoming climate neutral" and calls for making sustainability an integral part of the educational experience. Today nearly seven hundred colleges and universities have signed on to the commitment.[24]

Jean stands up from her desk and takes me to a cluttered cubicle where she introduces me to Karl Edelhoff, ASU's on-site solar project manager. It's Karl's responsibility to monitor all solar construction, keeping the installers on plan and on schedule. Once a solar array is up and running, he also has to make sure that all the equipment is in good working order.

Karl unplugs one of the university's Gem Polaris electric utility vehicles and we set off on a campus tour. Guiding this little open-air car along Tempe's streets and down campus alleyways, he calls out the university's solar highlights. We zip past PV-equipped libraries, parking lots, and classroom buildings, and he gestures toward the athletic center, where a solar thermal system heats the pool and shower water. He describes the circular solar array atop the Wells Fargo Arena dome—a visually striking landmark to planes approaching Phoenix's Sky Harbor International Airport. And he points to the just-completed solar canopy at Farrington Softball Stadium, offering its 1,500 spectators a reprieve from the blistering sun. The Arizona desert is great for solar, but it's brutal on Sun Devil fans when the softball season is at its peak in the late spring and early fall.

From the roof of Coor Hall (named after the university's fifteenth president, not the beer magnate), Karl identifies a few dozen solar landmarks as he scans the campus horizon. Most buildings have flat roofs well suited to solar, but he singles out several older roofs that should be replaced before solar panels are installed. Then he directs my attention to an imposing cylinder in shades of pink and tan stucco that looks like an amusement park carousel on steroids. This is Frank Lloyd Wright's Gammage Theater, he tells me, and it will never have solar panels on its roof. Tinkering with this architectural monument is taboo.

I refrain from asking Karl, an architect who trained at Syracuse University before heading west, what he thinks of Wright's late-in-life design. Instead we talk about his oversight of ASU's growing solar fleet. As most of ASU's solar arrays are owned and maintained by third-party developers, those companies bear the primary responsibility for maintaining their equipment and ensuring a consistent and reliable supply of power to the university. However, Karl has to make sure they respond quickly when breakdowns occur.

With solar panels purchased from a dozen solar manufacturers, ASU-Tempe has styled itself as something of a learning laboratory for photovoltaic technology. Some panels perform better than others, and at least one of the companies with panels installed on the campus—China's Suntech Power Holdings—is on the rebound from bankruptcy. The university has also experienced high rates of failure with one brand of inverters, the devices that convert direct current produced by solar panels into the alternating current that flows through most of the US grid.[25]

Even with these occasional pitfalls, solar power at ASU-Tempe delivers about 31 percent of the campus's peak electric load. System-wide, the university has more than 23 gigawatts of installed solar capacity on its campuses. In its use of solar power, it ranks second only to the University of Arizona, according to the Association for the Advancement of Sustainability in Higher Education. Rutgers, New Jersey's state university, ranks third with 17.4 megawatts of solar energy.[26]

On the Rutgers campus in Livingston, engineering professor Dunbar Birnie III meets me at one of the large solar-shaded parking lots that the university has recently built for some 3,500 of its car commuters. He arrives in a brightly painted blue-and-green Chevy Volt that looks like a NASCAR. Its side panels are emblazoned with the logos of its various sponsors: Rutgers Energy Institute, New Jersey Board of Public Utilities, and Recovery.gov—the grant-making arm of the American Recovery and Reinvestment Act. While Dunbar closely tracks the performance of the university's two electric vehicles under varied weather and traffic conditions, he focuses his primary research and teaching on solar cell design and fabrication.[27] Students learn about the science of photovoltaics, but they also delve into the efficiency and cost of different PV materials.

It's a hot midsummer morning, so Dunbar leads me to a welcome stretch of shade beneath one of the solar canopies. The canopies run

east-to-west, stretching the full length of the parking area. They are
braced at several-yard intervals by heavy steel I-beams. The southern
expanse of each long canopy lies nearly flat; the northern half slants
gently upward, as if hinged along a central spine. Together they create a gentle V, like an open, elongated book. As I look at the I-beams,
much more massive than the slender steel spans at ASU's solar parking lots, I wonder aloud: Is all this extra support needed to withstand
the weight of fallen snow? Dunbar says high winds are a much bigger
risk in this region, borne out by New Jersey's recent battering by Superstorm Sandy. The V-shaped configuration of the arrays, however,
is designed to keep accumulated snow from suddenly sliding off the
edges onto people walking to and from their cars.[28]

While New Jersey's winds may be stronger, its sun is weaker and
more intermittent than in places like Tempe, Arizona. Even so, Rutgers now draws over 60 percent of the power for its Livingston campus from eight megawatts of solar power on two parking lots plus
a smaller, 1.4-megawatt ground-mounted array. Thanks to a combination of state and federal incentives, the university's sustainability
coordinator, Michael Kornitas, expects good financial as well as environmental results in the immediate term. "We'll be pretty much cash-
positive right from the get-go," he predicted shortly after the second
solar parking canopy started generating power in 2013.[29]

Interested in what other colleges and universities are doing to bring
solar power to campus, I decided to see what, if anything, was happening at George Washington University in Washington, DC, where my
daughter Maya is a student. While solar collectors supply hot water to
a few dorms, I was disappointed to learn that solar photovoltaics have
not been incorporated into any of the big-budget buildings now going
up on GW's downtown campus. This includes the just-completed
LEED Gold-certified Science and Engineering Hall, which fills the
better part of a city block. The reason, I was told, is that the university, in maximizing the floor-to-area ratio in all new buildings, has
reached the District of Columbia's strictly enforced height limit and
can't accommodate even the few extra feet of rooftop elevation that
solar panels would require.[30]

On-campus photovoltaics may not play a role at GW, but the university is making a very substantial investment in off-campus solar. Together with American University and the George Washington University Hospital, it has committed to buy up all the power generated over a twenty-year period by three large solar plants now being built in North Carolina. GW's share of these projects' output is expected to cover more than half the university's electricity needs, eliminating an estimated 42,000 metric tons of carbon emissions per year—equivalent to taking 8,800 cars off the road. And the cost? Under the terms of the power purchase agreement signed with Duke Energy Renewables, the participating institutions will bear no up-front cost and will pay less for solar power than they have been paying for conventional electricity.[31]

Like Arizona State University, GW has joined the American College and University Presidents' Climate Commitment. Honoring this compact, the university administration has vowed to shrink GW's carbon footprint 40 percent by 2025, and to achieve carbon neutrality by 2040.[32] With its new solar power purchase, the university is taking a very substantial step toward meeting these goals.

On college and university campuses, student groups calling for fossil fuel divestment are an important new movement, awakening administrators and trustees to the environmental crisis at hand. If they succeed in ridding their school endowments of polluting and climate-destabilizing energy investments, they will have won half the battle. But what will replace the actual *use* of carbon-intensive resources by our 4,500 colleges and universities? The physical infrastructure and procurement practices of these institutions, as well as their stock portfolios, must lay the groundwork for a more sustainable future. Solar power can be a key player in bringing about this transformation.

—◊◊◊—

In addition to supporting private solar development through various incentives, the federal government is busy creating its own solar legacy. In March 2015 President Obama signed an executive order requiring all federal agencies to obtain 30 percent of their electricity from re-newable energy by 2025. He also called on federal agencies to rely on "clean energy sources" to meet 25 percent of their overall energy needs (electric and thermal) by the same date.[33] Since the federal govern-ment is the biggest energy user in the nation, these are momentous moves toward making America a lower-carbon economy.

Predictably, the Department of Defense accounts for the lion's share of federal energy use: about 80 percent of the government's over-all consumption.[34] Some progress has already been made in reforming this giant energy consumer's profile, set in motion by an earlier stat-ute calling for a military shift to renewable energy.[35] The motivation behind this statute, enacted in 2006, had more to do with protecting our military against energy vulnerability than promoting a cleaner energy portfolio.[36]

In a realm as expansive as national defense, translating raw percent-ages into renewable energy facts on the ground requires planning on multiple levels and across all branches of the armed services. While solar chargers and mobile solar arrays are being explored as ways to reduce troops' reliance on heavy batteries and precarious fuel convoys, the military's biggest solar potential lies at its fixed installations—some 500 worldwide with a total of 300,000 to 360,000 buildings.[37] At Davis-Monthan Air Force Base in Arizona, a 16-megawatt solar power plant supplies about 35 percent of the base's electricity, equiva-lent to meeting the needs of about 2,100 households.[38] Solar arrays at Naval Air Weapons Station China Lake, in California, will soon provide more than 30 percent of the base's yearly electricity load, sav-ing the Navy some $13 million in projected electricity costs over the next twenty years.[39] And in Georgia, PV arrays will be installed at three army bases by the end of 2016, supplying 18 percent of the army's statewide power needs.[40]

Solar power brings more than clean energy to US military bases; it also can provide a degree of independence from an electric grid that has proven all too vulnerable to natural disasters and human sabotage. In the wake of Superstorm Sandy, more than eight million households and businesses in the New York metropolitan area were cut off from electricity, some for more than a month.[41] Intentional attacks on the grid have also taken their toll, disrupting power to thousands of customers in rural Arkansas in the summer and fall of 2013, and knocking out local power lines as well as 911 service in San Jose, California, earlier that year. The attackers in these instances used old-fashioned methods of sabotage. In Arkansas, a fire was set at an electrical substation and a tractor was used to pull down electric poles. In San Jose, the vandals cut fiber-optic cables and knocked out seventeen transformers with their high-powered rifles.[42] According to the *Wall Street Journal*, there were "274 significant instances of vandalism or deliberate damage" to the American grid during a recent three-year period. And beyond these conventional attacks looms the specter of pervasive power outages caused by cyber terrorism.[43]

Acutely aware of the risks posed by over-reliance on power sources outside their bases, defense planners are looking toward a future when on-site renewable energy, combined with battery storage and backup diesel or gas-fired generation, can secure uninterrupted power to military installations at home and abroad. One of these microgrids has already been launched on a pilot scale at Fort Bliss, an army base in Texas, and others are in various stages of development.[44] "Resilience" is the term often used to describe this new shift toward energy survivability in military as well as civilian communities.

—⁓—

From sports arenas and commercial buildings to university campuses and military bases, solar power is finding its way into our built environment on a scale that can substantially shrink the carbon footprint of many of America's bigger institutions and enterprises. Yet we have hardly begun to scratch the surface of this renewable energy gold mine.

Look out the window the next time you're landing on a flight into any American city. Observe the rooftops of big box stores, bare but for the occasional skylight or air-handling unit. Take note of the countless parking lots that engulf so many of our shopping centers and corporate office parks. Scan the factories and warehouses with their colossal roofscapes. Not all of them may be well-suited to solar, but pioneers like Tom Gros, Richard Bass, Kathy Doyle, Alan Epstein, and the Avidans can broaden our awareness of the not-so-hidden private sector prospects that have escaped our attention. Our eyes are also being opened to the steps some of our colleges and universities, and even our defense installations, are taking to promote a cleaner energy future. In all of these instances, far-sighted leadership can inspire us to push beyond complacency and embrace sustainable energy opportunities that are both technologically feasible and fiscally sound.

CHAPTER THREE

Local Communities Capture the Sun

HARNESSING SOLAR ENERGY'S FULL potential will require millions of people making smart decisions about the ways they power their homes and businesses. Those smart decisions are seldom made in a vacuum, however. While federal tax credits and favorable state policies are key catalysts to solar investment, local governments are at the forefront of solar power pioneering. By adopting solar-friendly building codes and zoning ordinances, they can ensure that solar arrays are embraced as a necessary and welcome part of the contemporary urban landscape. Through tax abatements and other financial incentives, they can help bring solar investments within easier reach. And by expediting permit procedures, they can make sure that local solar adopters don't have to wait weeks, or even months, to get their solar systems approved.

Beyond helping constituents step up their reliance on solar power, many cities and towns are leading by example, installing solar arrays on municipal buildings, public schools, and other government properties. Some have even dared to break away from the giant power companies that have long supplied their electricity, inviting local citizens to buy their power instead from community-run utilities that set ambitious goals for renewable energy use. While worries about the health impacts of fossil fuel-burning are a common motivation, local political leaders and the citizens who elect them make it clear that they have a broader global agenda as well: to play their part in the battle against global warming.

Lancaster, in California's Mojave Desert, has a population of 160,000 and a nasty legacy of gang violence. In recent years, the city's image has been radically changed by Mayor R. Rex Parris, a class-action lawyer with unbridled determination, enough personal wealth to eschew campaign contributions, and a penchant for publicity. When first elected in 2008, he made crime his number-one target. A hotel that hosted gang events was shut down, a Cessna was dispatched daily to conduct aerial surveillance of tough neighborhoods, and pit bulls were leashed and neutered. Over a four-year period, the mayor claims that 20,000 youth ended up behind bars. Without a hint of hesitation, he looks me in the eye and says: "I don't care about the Constitution once you join a gang." His grandfatherly white beard, rosy cheeks, and ready smile only partially defuse the harshness of his words.

An unapologetic law-and-order politician, Parris is equally unabashed in his call for an all-out war on climate change. His environmental concerns date back to his student days in the 1970s, when ozone depletion and smog were the hot issues. He remembers working at a gas station and arguing with his boss about the air pollutants that seemed to be mounting inexorably, clouding California's skies and polluting people's lungs. He asked his boss: "What about your grandchildren?" The response he got shocked him: "If they have to wear gas masks, then they just have to wear gas masks."

It took quite a number of years for global warming to emerge as a widely acknowledged menace, but when it did, Parris was primed. Today his apocalyptic view of climate change is matched by a deep respect for the innovators who are coming up with ways to avert its full devastation. "The people who are now getting attracted to alternative energy are probably your brightest people in the world," he asserts. "They are the ones who can see the end, and the end game is horrible."

A staunch political conservative with close ties to the Christian Right, Lancaster's top elected official has angered his colleagues in the Republican mainstream by questioning the motives underlying their refusal to treat climate change as an urgent threat. "Who supports the Republican Party *really?*" he asks me. "It's not the churches. It's not conservative America. It's the *oil* companies." He deplores the political jockeying that has come to polarize debate on this issue. "The moment you say 'Republican' or 'Democrat,' it's now in the political realm and you've lost any hope of moving your agenda forward."

Parris had already begun pressing his own agenda for Lancaster's solar future when he attended an economic forum in Shenzhen, China, in 2010. But it was there that he staked out a characteristically bold stance for his city. "We were sitting at a table that had to be the size of this building," he says, his hand gesturing toward the suite of city hall offices outside the conference room where we have gathered with several of his aides. One after another, the global dignitaries read their prepared speeches, and Parris realized his turn was coming. What could this "kid from Lancaster" say to make his mark? His feverish note-taking bewildered another member of the Lancaster delegation. "It's *not* that interesting, Rex!" his neighbor whispered, thinking he was writing down what others were saying. The mayor ignored him: he was about to declare his city "the solar capital of the world."[1]

Since then, Parris has rallied his staff behind a multipronged effort to deliver on his brash proclamation, and today he's at least partway there. Though Lancaster ranks thirty-first in population among California cities, it holds fifth place in total installed solar power. By the end of 2014, it had 25 megawatts of photovoltaic capacity in place—enough to meet the needs of more than 5,300 average California households.[2] Solar arrays are on the city hall roof, and they cover a large adjacent parking lot too. They also have been installed on twenty-five public school campuses, the city's football stadium, and its performing arts center. Reaching beyond the municipal sphere, the mayor has encouraged local businesses to install PV panels—for example, the Toyota dealership where I saw an entire fleet of new cars shaded beneath several rows of steel-frame solar canopies.

Heather Swan, senior projects coordinator for the newly created Lancaster Power Authority, drives me to Amargosa Middle School to see the solar arrays that were recently installed atop parking areas, shading playground space, and covering walkways to athletic fields. This school, like others in the city, had previously been hit with giant electric bills because so much of its power usage occurred during hot daytime hours when Southern California Edison was charging peak rates. Rather than having Armargosa and other schools purchase the solar arrays outright, the Lancaster Power Authority has financed their installation while SolarCity owns and operates the plants. The overall cost of these projects has been substantially reduced because SolarCity, as a private, for-profit company, can take advantage of the 30 percent federal renewable energy investment tax credit. The municipal utility, in turn, charges the schools 12.5 cents per kilowatt-hour for the power produced on their campuses—far below the 19 cents per kilowatt-hour that they had been paying, on average, to Southern California Edison.[3] Annual savings on the school system's electric bills are estimated at $325,000.[4]

Mayor Parris's more pathbreaking move was to mandate that all new housing incorporate solar power. Effective January 1, 2014, every new residence in Lancaster must be matched with at least one kilowatt of solar generating capacity. Builders can meet this threshold—enough to meet a fifth to a quarter of the average California household's electricity needs—by installing one-kilowatt solar arrays on the roof of every home in a subdivision, or they can build larger arrays on a smaller number of homes so long as the total installed capacity meets the cumulative obligation.[5] For new multifamily dwellings, PV arrays can be mounted on parking lots, shared rooftops, or adjacent land.[6]

With housing demand in a slump since the recession of 2008, Lancaster isn't seeing much new residential development. This means that the city's residential solar mandate—modest but inspiring—will not lend much momentum to the mayor's proclaimed goal of making Lancaster "a zero net energy city." When I mention this, the mayor's response is brusque: "Give me time." He admits that zero net energy *neighborhoods* may be hard to achieve in the short term, but he quickly adds: "The zero net *city* is not a problem for me." By 2016, when his third term as mayor runs out, he expects to have achieved his aim.

Jocelyn Swain, a planner with the city, helps me understand how she and her colleagues define zero net energy. The definition, like so much about Mayor Parris's governing style, strays from orthodox interpretation or application. To most environmental planners, zero net energy means that the net energy *used* by a building, a company, or in this case a city, nets out to carbon neutrality through energy efficiency measures, demand management, and/or the use of non-emitting energy resources like the sun. Yet Jocelyn tells me that, in addition to counting the solar electricity that is generated and used within Lancaster, the city's zero net energy balance will include solar power that is locally produced *but sold to outside customers* such as neighboring municipal utilities and even investor-owned utilities like Southern California Edison. Those utilities will presumably credit the solar electricity that they purchase toward their own renewable energy obligations under California's renewable electricity standard. When I suggest that this seems to invite double-counting by multiple entities eager to claim credit for the same renewable energy, she acknowledges this could happen "in theory."[7] Then, in an effort to reassure me, she shifts our focus away from who is buying the electricity to where the solar-generated electrons are actually being used. Because the solar electricity is being fed into the grid at a transmission substation in or near Lancaster, she tells me it's likely that "someone in the local area is using it." Technically she may be right, but the double-counting relates to the electricity marketplace, not to the physics of power transmission.

Lancaster's framing of its zero net energy goal may be flawed, but the city's dedication to moving solar power forward on a prodigious scale is beyond question. Working with Jocelyn and her colleagues are energy developers who are equally eager to take advantage of Southern California's solar-friendly climate and lucrative power market. One key investor, Sustainable Power Group (sPower), is a Utah-based firm with solar, wind, hydro, and biomass projects in various stages of development across much of the West. Already sPower has signed power purchase agreements for two utility-scale solar installations in Lancaster; the electricity will be sold to the cities of Pasadena, Riverside, and Azusa.[8] Additional sPower solar plants are moving quickly through Lancaster's permitting pipeline.

Lancaster's eagerness to expedite new solar development is a great match for sPower's portfolio, which prioritizes projects sited on previously disturbed properties, minimizing the environmental conflicts that can arise when solar farms are built on pristine lands. Most of the utility-scale PV projects that the city has permitted are on privately owned farmland where alfalfa, carrots, and onions used to grow. Just a few of the proposals are on city-owned land: a defunct fairground and a former school property. All are privately developed projects that will sell their power to entities outside Lancaster.

Given all the technical and logistical hurdles involved in luring private solar developers to Lancaster, I ask Jocelyn: "What does the city get out of this?" In addition to the reputational benefits of Lancaster becoming—by its own unique definition—zero net energy, she becomes particularly animated when talking about the hundreds of jobs that are created by new solar projects. Even short-term construction work is a boon to the community, and to the people who land those jobs. As she speaks, I find myself thinking back to all those residents jailed during the mayor's assault on crime. Quietly, I wonder if this surge in solar employment has kept at least a few Lancaster citizens from ending up behind bars.

Jocelyn then speaks of the city's revenue gains. On one hand, property taxes don't increase because California law prohibits renewable energy projects, including PV installations, from being assessed as taxable real estate.[9] On the other hand, Lancaster enjoys a big boost in sales tax receipts on all the equipment and services provided to new solar projects. Designating Lancaster as the point of sale for these installations ensures a substantial flow of revenues to the city. "When you're talking about hundreds of thousands of panels," Jocelyn says of utility-scale projects, "it's a one-time deal, but it's still a revenue source." To a lesser degree, sales tax revenues will provide an ongoing benefit to the city as project owners spend money to maintain their facilities in the years ahead.[10]

Mayor Parris is lucky to have smart professionals like Jocelyn Swain and Heather Swan to ground his ambitions in careful planning and implementation. *They* can worry about making the mousetrap work while *he* trumpets his bold dreams. Those dreams, I learned, are fueled by proud memories of the military aircraft industry that once thrived in the area. "We saved this planet once before," he says. "We did the B-1, the B-2, the F-100—you name it. . . . If it flew, it came out of here." His nostalgia then yields momentarily to less happy recollections of his father's place in that tumultuous Cold War constellation. He recounts his father's sporadic work on an aircraft factory production line. "One day we were rich, the next day my dad was on unemployment."

Ever the optimist, Parris quickly brushes these recollections aside. Solar energy, somehow, offers a less ephemeral promise than the boom-bust history of the local aircraft industry. "We will become the center of the universe when it comes to solar," he glows. "*That's* what I want."

—⁊⁊⁊—

Marin County occupies a lush peninsula stretching northward from the Golden Gate Bridge. Its median household income tops $90,000, almost twice that of Lancaster, and its strongly progressive tilt got the senior George Bush in trouble when he called Taliban fighter John Walker Lindh—who spent part of his childhood there—a "misguided Marin County hot-tubber."[11] A long way from Lancaster geographically and culturally, it is a more predictable home to environmental enlightenment.

Sky-high land values are part of the reason why Marin County has headed down a clean energy path quite distinct from Lancaster's. Instead of becoming a net solar exporter, Marin has primarily focused on lightening the carbon footprint of the energy it brings *into* the community.

In leafy San Rafael, Marin's county seat, I met with Dawn Weisz, a clean energy maverick every bit as determined as Rex Parris, but with a soft-spoken, methodical demeanor wholly different from the Lancaster mayor's swashbuckling bravado. Dawn is the executive officer at Marin Clean Energy, a pioneering nonprofit that has taken advantage of AB 117, a state law passed in 2002 allowing communities to procure, or "aggregate," their own electric supplies for local customers, instead of buying their power from the state's big investor-owned utilities.[12] This new approach is called Community Choice Aggregation (CCA), and where it is adopted, the investor-owned utilities

remain responsible for transmitting and distributing the locally pro-
cured electricity at regulated rates comparable to those charged to its
regular customers. Similar laws have been adopted in five other states:
Illinois, Massachusetts, New Jersey, Ohio, and Rhode Island.

Shortly after AB 117 became law, activists in the town of Fairfax
began to press the Marin County Board of Supervisors to consider
forming its own power-purchasing entity. The proposed goals for this
body would be to increase the use of green energy, promote energy efficiency, and create local economic benefits. Years of studies and public
polling ensued, and in 2008, eight local jurisdictions voted to break off
from Pacific Gas & Electric (PG&E), the giant investor-owned utility
that sold Marin County its power and still serves most of northern
and central California. Instead, they formed a local power authority
now called Marin Clean Energy.[13]

By the time Marin Clean Energy (MCE) was created, California
had already adopted the nation's most aggressive renewable portfolio
standard, requiring PG&E and other utilities to supply 20 percent
of their power from renewable energy sources by 2010, rising to 33
percent of total delivered electricity by 2020. Yet Dawn and her colleagues were committed to moving Marin County far more quickly
toward a low-carbon energy future. They created two options for customers. Under a Light Green program, customers would get at least
25 percent of their electricity from renewable energy, later raised to 50
percent. Deep Green subscribers would be charged a slightly higher
rate to receive 100 percent of their power from renewable sources.

The initial rollout of Marin Clean Energy was very rocky. Unhappy
with the advent of this renegade local power provider, PG&E spent
$4.2 million trying to undermine its work.[14] Telemarketers saturated
the area just as the program was being launched, urging customers
to opt out and remain with PG&E. Mailed notices warned customers that they would be subjected to higher electric rates and bigger
financial and operational risks if they chose MCE, with its unproven
track record. The California Public Utilities Commission came down
quickly and firmly on PG&E, chastising it for "acting in a deliberate
manner to subvert the plain meaning of AB 117," which requires utili-

ties to cooperate fully with community choice aggregators, rather than spending ratepayer dollars fighting them.[15]

Dawn remains outwardly calm as she reflects back on those early tumultuous days. "I think they saw this as a competitive threat to their monopoly position," she says. "PG&E was meeting with decision makers throughout Marin saying, 'This is risky, their prices are going to be twice as high, their products won't be as green, there's not enough renewable energy out there.'" That last claim she finds particularly ironic. "I can't believe how much power we get offered on a routine basis," she says. "There's a *lot* of renewable power out there."[16]

Despite the utility's ongoing interference, Dawn and her colleagues at MCE pressed on. "We just kept our eye on the prize and kept going," she recalls. Today MCE is a striking success: it has 125,000 subscribers, or about 80 percent of the customer base in the 13 towns and cities that now participate in the program.[17] Its 51 percent renewable resource mix reportedly outpaces every other utility in the state.[18] While PG&E claims its "carbon-free" power also amounts to 51 percent of its portfolio, Dawn is quick to clarify what this means.[19] "Renewable power and carbon-free power aren't one and the same." PG&E, she explains, is counting its 22 percent reliance on nuclear power as "carbon-free." It is also including the electricity it gets from large-scale hydro dams, which are disqualified under California's renewable portfolio standard, largely because of the environmental controversies surrounding big dams.[20] "We're aiming to be 55 to 60 percent carbon-free," she says, "but we're doing that without buying any nukes or large hydro." Eventually, she adds, MCE's goal is to have a 100 percent renewable energy portfolio.

And the costs? Dawn claims that Light Green customers pay rates comparable to PG&E's. Sometimes the Light Green rates are slightly higher; sometimes they're somewhat lower than PG&E rates. Deep Green enrollees pay a premium of five to six dollars per month for their all-renewable energy portfolio.

Solar energy remains a fairly modest contributor to MCE's energy mix—about 6 percent, compared to wind's 30 percent. Most of those solar-powered electrons come from PV facilities inside California but outside Marin County. This is explained, in part, by geography: large-scale solar power plants are mainly being built in southern California where "insolation," or solar radiation, is much higher than in Marin's temperate climate. Marin's super-pricey real estate market also makes it hard for solar developers to find good local sites for large solar projects. Not surprisingly, then, the real growth in solar power within Marin County is on residential and commercial properties. Households and businesses with PV arrays qualify for MCE's net energy metering tariff, a one-cent-per-kilowatt-hour credit for any surplus power they generate during a monthly billing cycle. That extra penny per kilowatt-hour is added to the rate that net-metered solar customers in California are generally entitled to.[21]

Beyond compensating homes and businesses for surplus solar generation, MCE is also luring companies to build solar power plants within the county through a "feed-in tariff," or a fixed rate guaranteed over a twenty-year period, for all the electricity they generate. One such plant is already operating at San Rafael Airport, and others are in various stages of planning. Dawn speculates on the kinds of operations that might qualify for the feed-in tariff. "It could be a dairy. It could be a sewage treatment facility, a landfill, or anyone who has enough space to put up a lot of solar."

Other jurisdictions within PG&E's service territory have had mixed results in their attempts to break loose of the utility giant. In 2007 a number of communities in the Greater Fresno area took initial steps to create the San Joaquin Valley Power Authority.[22] PG&E spent close to $4 million fighting this effort, and the initiative languished. In San Francisco, PG&E shelled out $1.6 million to beat down a CCA program called CleanPowerSF.[23] After years of acrimonious debate, Mayor Ed Lee finally ordered city officials to offer this plan to local electric customers by late 2015.[24] And in Sonoma County, just north of Marin, community choice aggregation has prevailed. As in Marin, Sonoma Clean Power offers ratepayers two choices: CleanStart, based on 33 percent renewable energy at a rate projected to be 5 to 8 percent cheaper than PG&E's, or EverGreen, representing 100 percent renewable energy at a 20 percent higher rate than CleanStart power.[25] Early results suggest that 80 to 90 percent of Sonoma County ratepayers will end up joining Sonoma Clean Power, though only a small fraction are likely to sign up for the more costly EverGreen option.[26]

If the California experience is any indication, wresting clean energy customers from powerful utility monopolies requires utter determination laced with a healthy dose of idealistic vision. Legislation favorable to community choice aggregation and the California Public Utilities Commission's support have also been critical. Through years of struggle, Dawn Weisz and her Marin Clean Energy colleagues have created a new model for community autonomy—one that allows local citizens to press beyond state-set goals for low-carbon power. But theirs is not the only path to local energy self-determination.

—⁓—

California cities with their own municipal utilities, like Palo Alto, have had a much easier time pursuing sustainable energy goals than jurisdictions that are within the service territories of utility monopolies like PG&E. The City of Palo Alto Utilities (CPAU) has been the local electricity provider since 1900, and its role as renewable energy innovator goes back to 1980, when it offered low-interest loans for solar water-heating systems. In 1999 CPAU began offering rebates to purchasers of solar PV systems, and a few years later it established a renewable electricity standard that outpaced the state by half a decade, committing the utility to a 33 percent renewable energy mix by 2015. Then, in 2013, the city council ratified a plan to make CPAU's portfolio 100 percent carbon-neutral, effective immediately. A snappy rap video on the city's website celebrates this move. With shots of wind turbines, a hydro sluiceway, and PV arrays flashing in the background, a cool kid wearing sunglasses and a hoodie struts across the screen chanting about the city's passion for carbon neutrality. "Climate's changing so let's face it. Let's FACE it!" he exhorts."[27]

The steps being taken by Palo Alto go way beyond hip-hop lyrics. Through power purchase agreements signed in 2013, CPAU will own the output of three solar PV installations, now being built on idle farmland in central and southern California. By 2017, when these projects are completed, they will meet about 18 percent of local power demand. Together with other renewable energy contracts, the utility will be getting 48 percent of its electricity from carbon-free sources that qualify under California's renewable portfolio standard. Most of Palo Alto's remaining electricity needs will be met imperfectly, from an environmental standpoint, by large hydro projects located in the state. The incremental cost to local ratepayers for achieving carbon neutrality is expected to be about 0.11 cents per kilowatt-hour or about $3 per year for the average residential customer.[28]

In San Francisco, progress toward community choice aggregation has been slow, but solar is taking hold in other ways. The San Francisco Public Utilities Commission is not just a local regulatory body; it is also the city's power company, providing electricity to public buildings and operations ranging from local public transit to water supply and sewage treatment. In all, it meets about 17 percent of local power demand, and a growing though still-modest share of that electricity comes from the sun.[29] Solar arrays are now operating at the Moscone Convention Center, San Francisco International Airport, three of the city's water treatment plants, and more than half a dozen other city-owned properties. Most stunning, though, is the solar installation that caps the Sunset Reservoir on the often-fogbound western rim of the city. When I asked Pasha, a garrulous taxi driver, to take me there, he asked me why. I told him and he laughed. "I used to live in that area. It's foggy at least three hundred days a year!"

Luckily, I happened to visit the reservoir on an afternoon when the sky was deep blue and the sun was bearing down hard. To this one-time visitor, it looked like the perfect place for one of the nation's largest urban solar installations. Surrounded on all sides by neat stucco homes, the reservoir is enclosed beneath a concrete roof that covers an area nearly as large as twelve football fields. Tens of thousands of solar panels are now mounted on this colossal concrete surface. What a great use for an area otherwise off-bounds to the public in a security-minded era! I peered at this spectacle through a heavy chain link fence, and beyond it saw the Marin County headlands across a clear expanse of Pacific coastal waters.

The Sunset Reservoir plant and other solar installations supply less than 1 percent of the San Francisco PUC's electric needs, but they offer an appealing counterpoint to the PUC's primary source of power: the controversial Hetch Hetchy hydro complex. For decades, conservationists have decried the damage these dams have done to wooded canyons on the western edge of Yosemite, yet the sale of surplus power from Hetch Hetchy has provided a key revenue stream for San Francisco's Go Solar incentive program. According to the utility commission's communications manager Charles Sheehan, these funds have facilitated the installation of more than 6 megawatts of solar power on San Francisco businesses and homes, including low-income housing and multi-unit apartment buildings.[30]

One of the more unusual renewable energy initiatives coming out of the Bay Area uses online marketing to crowdsource solar power on buildings whose owners—typically small nonprofits—would otherwise find these clean energy investments beyond reach. Billy Parish and his team at Oakland-based Mosaic began soliciting online investors for solar projects in January 2013, and by September of that year, they had brought in over $4 million from 2,000 participants in 44 states.[31] In one eight-hour period, they raised $225,000 from 335 people for two PV installations at an affordable housing complex. The average investment was $670.[32] Mosaic offered its first loans to solar developers at about 5.5 percent and promised its investors a 4.5 percent return on their money—much better than ten-year Treasury notes, as one reporter observed.[33] Whether this model can be ramped up substantially remains to be seen, but in April 2013, Mosaic received a permit from the California Department of Corporations allowing it to issue up to $100 million in securities to investors.[34]

—◊◊◊—

While California remains the nation's leader in solar innovation and investment, a number of East Coast cities and towns are finding their own ways to step up the pace of solar development in their communities. One of these is New Bedford, a city in southeastern Massachusetts built on whaling, and then textiles, that is struggling to find a foothold in the twenty-first century. Though it has the highest-value fishery in America, the city's eroding industrial base has left thousands jobless. Unemployment levels topped 13 percent in 2013.

Matthew Morrissey's New Bedford lineage goes back five generations. After a successful stint with a software company in the late 1990s, he returned to his New Bedford roots and, in 2005, ran for mayor at age 31. Though he lost, he was soon asked to head up the New Bedford Economic Development Council, where he spearheaded a comprehensive assessment of the city's existing economic base and future prospects. "Cities like ours often don't look authentically at their assets and their challenges," he observes. "Where's the private market going? Where's federal and state policy going? Where is the venture capital going? Where are the challenges in the workforce or in land use?"[35]

Some two hundred community leaders—business executives, neighborhood organizers, labor representatives, and educators—participated in this process. Energy emerged as a key potential driver of economic growth, with a particular emphasis on offshore wind. For a New England port city like New Bedford, offshore wind looked especially promising, not just because the Cape Wind project in Nantucket Sound was moving toward final approval, but because proposals were emerging for major wind development all along the Mid-to-North Atlantic seaboard. New Bedford would soon start positioning itself to become the hub for building and servicing offshore wind farms up and down the coast.

Before Matt Morrissey changed jobs to become director of New Bedford's Wind Energy Center, he presided over a multiyear effort to develop the city's solar energy potential. He did so with the mayor's blessing, but his primary partner in framing this new enterprise was a seasoned environmental policy maker named John DeVillars, founder of an environmental consulting firm called BlueWave Strategies and, later, a renewable energy development group called BlueWave Capital. BlueWave advised New Bedford and several other Massachusetts municipalities that had received grants under the Energy Efficiency and Conservation Block Grant Program, an Obama administration clean energy initiative funded by the American Recovery and Reinvestment Act.

New Bedford used its block grant, in part, to launch a program to clean up the municipal government's electricity supply by installing up to 20 megawatts of solar power.[36] To catalyze this effort, BlueWave set about identifying specific sites that would be well suited to solar. First they surveyed city-owned properties and found a handful of roofs—three schools, a municipal agency, a gym—that had the right solar orientation and were in good enough shape to host solar arrays. They also identified two capped landfills, one of which was adjacent to the local high school. But finding enough real estate within the city limits proved difficult, so they also searched the surrounding countryside for suitable properties. "There's no way we could have even gotten to 10 megawatts without going outside the city," DeVillars says.

One promising prospect BlueWave came up with was a stretch of scrubby woodland adjacent to Little Quittacas Pond, the main reservoir for New Bedford, located several miles north of the city. When I drove to the area, I immediately understood why Matt Morrissey calls it the "dead pine site." If he hadn't given me that cue, I might not have spotted the array at all, discreetly tucked behind a landscaped berm along a ramrod-straight country road with small suburban-style ranch houses on one side and a large, desiccated stand of pine trees on the other. The array's fourteen thousand panels are expected to supply 100 percent of the power needs of the water-pumping and treatment facilities at Quittacas, at a fixed cost of eleven cents per kilowatt-hour over twenty years, well below the fifteen to sixteen cents per kilowatt-hour paid by the municipality in recent years. As the city's public water supply ranks second only to sewage treatment as a consumer of electricity, this project's fiscal benefits to New Bedford are momentous.[37]

The BlueWave team also came upon a privately owned sand-and-gravel pit in Plymouth, some 40 miles away. They knew that a solar installation at this location would count toward the city's net-metered power supply because it was within the "load zone," or contiguous service territory, of the local electric distributor, NSTAR. Most of the electrons will never actually make their way to New Bedford, but under Massachusetts law they can be treated just as if they were generated within the city limits. "Virtual net metering" is the term policy makers use to describe this sort of off-site supply of power to the grid.

"There are so many pieces to this business," DeVillars observes. "If any one of them goes wrong, you don't have a project."[38] As a longtime environmental regulator, he has a keen eye out for permitting problems that could end up killing a project. A BlueWave real estate specialist makes sure landowners are willing to lease or sell their property, and another team member verifies that there is available capacity on nearby power lines. If all those pieces fit into place, a financial expert sets about finding a willing investor that can make a reasonable return selling power while benefiting from the 30 percent federal investment tax credit, the five-year accelerated depreciation allowed under federal tax law, and the sale of SRECs for every megawatt of electricity produced by the project.[39]

Finally, someone has to get the equipment installed and make sure it runs properly. For several of the New Bedford projects, that key role has been played by a century-old sign-company-turned-solar-integrator. When Phil Cavallo bought Beaumont Sign Company in 2006, it was a frugal operation run by the family's third generation. Its customers ranged from chain stores to area banks, and its revenues were about $2 million per year. A native of greater Philadelphia who had logged in nearly three decades as a self-described serial entrepreneur, Phil had done everything from designing missile guidance systems for Raytheon to launching telecom start-ups. It was time, he felt, to go local. "I had flown all over the world doing business for big corporate America," he tells me. "I wanted more of a community-based business where I could give back in ways and develop a relationship in ways that I hadn't been able to do in any other job."

Phil's initial intention was just to keep on selling big commercial signs. It only took a few months, though, for his entrepreneurial drive to kick in. He looked at Beaumont Sign's 1,800 loyal customers and said to himself, "There's something else we can sell to those customers besides signs." Initially he thought of small-scale wind turbines. Turning to a colleague at Beaumont, he asked: "Hey, is there a big difference between putting up a wind turbine and putting up a billboard? It's got a foundation, it's got a tower, it's got a sign at the top." The colleague was OK with the idea, so they gave it a try. After being hired by the State to take down twenty-two turbines that were improperly permitted, they set about building some new ones, but the slow pace of moving wind proposals through local planning bodies sapped their ardor.

Phil then turned his focus to solar. Reading *Inc.* magazine, he was impressed to see that PayPal tycoon, Tesla creator, and SolarCity bankroller Elon Musk had been named 2007 Entrepreneur of the Year.[40] What really struck him was the scuttlebutt that SolarCity might go public, even though it had only been launched the previous year. "From VC [venture capital] and working in the Valley, I knew that, to go public, you need to have an unfair competitive advantage, you have to be in discontinuous innovation, and you have to have huge market cap potential."

Phil's vernacular swings sharply between the heady lingo of a venture capitalist and the down-home jargon of a community entrepreneur. One moment he's back "in the Valley." The next he's in gritty, workaday New Bedford. "I'm lookin' at this install business, and I'm goin': 'Hey, we can install signs on rooftops. We can do this all day long—laggin' 'em into roofs. Why can't we install solar panels?'" He lost no time jumping into this new field. In a training program for solar installers offered by a leading PV manufacturer, he remembers a moment when one of his classmates quipped: "You know the definition of a successful solar company?" The guy paused for effect before answering his own question: "An electrical engineer with a backhoe." As a trained electrical engineer with a yard full of sign-installation equipment, Phil had a somewhat different response: "I got a crane truck—*better* than a backhoe!"

Starting small, Phil sent Beaumont crews to install a few arrays on residential rooftops in neighboring Dartmouth, Massachusetts. He calls this the "off-Broadway performance scenario." Soon enough, he had twenty-plus home solar contracts on his desk, and his new line of business was drawing the attention of the local media. It was the right moment to make the leap to larger projects, and to give the company a new name: Beaumont Solar. The timing was perfect, as BlueWave's DeVillars and Morrissey were ready to move New Bedford's solar initiative forward. The transformed sign company landed several plum jobs.

One of Beaumont's breakthrough projects was a 500-kilowatt PV facility just across the street from New Bedford High School. Like the school itself, the solar array was built on a closed, capped landfill. It now supplies a quarter to a third of the high school's electricity, but Phil says this may increase if the building's roof passes muster as a secure spot for additional panels. We peer through a chain-link fence at several dozen orderly rows of panels before heading out to Sullivan's Ledge, a former waste dump so polluted that it remains an actively treated, EPA-designated Superfund site.

Once a quarry, Sullivan's Ledge was used by local industries through much of the twentieth century as a dumping ground for waste oil, PCB-laden electrical capacitors, other volatile chemicals, and demolition materials. It was fenced off in 1985, but it wasn't fully capped for another fifteen years. Even today, an on-site treatment plant pumps and decontaminates groundwater while a network of vent stacks channels the release of landfill gases. When Phil and I arrive, a specially trained crew is ever-so-gingerly preparing the earthworks for a 2-megawatt PV array. We walk out onto a muddy stretch of land where two guys operating miniature excavators are carefully digging a trough for the electrical cabling. To protect the clay contaminant barrier that is buried just 36 inches underground, there are strict weight limits—5 pounds per square inch—on all equipment allowed at the site. The excavator operators know that their small, smooth-edged shovels can dig no deeper than 18 inches as they work the soil.

The first few hundred ground mounts for solar panels are already in place along a stretch of flat land abutting a cluster of self-storage sheds and a gas station. Hefty, above-ground concrete blocks hold triangular steel braces in place—a safer mounting technique for panels than driving steel poles deep into the ground given the site's pollution hazards. Ultimately this facility, with its eight thousand PV panels,

Phil Cavallo, president of Beaumont Solar, with concrete-ballasted
mounts for solar arrays at Sullivan's Ledge Superfund site,
New Bedford, Massachusetts. (Photo by author)

will save New Bedford about $2 million in electricity costs over a
twenty-year period.[41]

In all, New Bedford's solar initiative is expected to supply about
60 percent of the municipal government's power needs, reducing its
electric bills by about $1 million annually.[42] Better still, the city has
incurred no capital costs because the solar installations are all owned
and operated by a third party. Power purchase agreements with companies like SunEdison and Con Edison Solutions set fixed electric
rates that are well below what the city has been paying for its power.
And on top of those financial benefits are the local jobs at firms like
Beaumont Solar. Fifty people are employed at Beaumont today, and
quite a number of them have come straight out of the vocational education program at New Bedford High. Beaumont has paid for others
to attend night school, to get trained as licensed electricians. Several
have come from troubled homes. "It's sort of a shot in the arm to get
these kids feeling good about themselves," says Phil.[43]

—ɯ—

This old New England whaling town's solar initiative may not rival sun-saturated Lancaster's dream of becoming "solar capital of the universe," nor will it match the green energy strides taken by Marin Clean Energy or the carbon neutrality already achieved by the City of Palo Alto Utilities. Yet all these efforts are emblematic of towns, cities, and counties across the country whose citizens and political leaders are seeking new local pathways to energy sustainability. America has nearly forty thousand general-purpose local governments; each can be a laboratory for energy renewal.[44]

As I learned from my travels, enormous opportunities for energy transformation lie far beyond the safe and familiar boundaries of America's liberal enclaves. Red and blue stereotypes are a hindrance, not a help, if we are to draw upon the ingenuity that resides in so many localities across the nation. The hubris of Rex Parris is as valuable as Dawn Weisz's studied determination to move us closer to energy sustainability. Daring leadership, farsighted policies, and an engaged public are key ingredients in moving communities large and small beyond the constraints of carbon-based energy resources.

CHAPTER FOUR

Wastelands Redeemed

JILL BARONE WARNS ME that she may become Bedminster's Erin Brockovich if KDC Solar is allowed to move ahead with its plans for a stretch of semi-idle farmland adjacent to her home. "This is my sanctuary," she says as we walk past her white clapboard rambler and pause by the koi pond, complete with cascading waterfall, lily pads, stone crocodile, and, of course, colorful fish. A miniature windmill—for decoration only—stands on a nearby hillock. We then move on to her swimming pool—small but meticulously landscaped. "Jilly's Pool" is displayed on a lifesaving ring hung near the water's edge. "I dreamed about a house like this and was blessed to have it, so I'm devastated."

By local standards, Jill's three-acre spread is a modest one. Zoning in this upscale New Jersey community was boosted to ten acres per residential unit after a mixed-income housing development raised the hackles of many old-timers when it was built a few decades ago. Famed for fox hunting, polo, and horse breeding, Bedminster is as close to aristocratic as it gets in the Garden State. Jacqueline Kennedy Onassis once owned a horse farm there; so did King Hassan II of Morocco. Forbes family members are still prominent in the community, recently joined by Donald Trump and his eponymous Trump National Golf Club with its forbidding fieldstone walls, guardhouse, and prominently displayed family crest—the one that created such a stir with the Scottish heraldic authorities because Trump hadn't bothered to get their approval.

Logic might have led Alan Epstein and his colleagues at KDC Solar to look elsewhere for a site to build a 10-megawatt PV facility, but the Kirby Farm had a few apparent advantages: it had long ago fallen into disuse; it was just across the highway from a major pharmaceutical company's US headquarters; and it was only a few miles from KDC's corporate offices. Epstein's team had already completed several solar installations on open lands in New Jersey, including a PV array on a cornfield that now supplies the Lawrenceville School, a prep school just south of Princeton, with 90 percent of its power. Another of its facilities, the Middlesex Apple Orchard, provides 100 percent of the electricity for a multi-building county correctional complex.[1] Epstein viewed the 110-acre Kirby Farm as a prime prospect for his firm's next big ground-mounted solar array. "That particular parcel of land was left lying fallow for years," he asserts. "There's nothing going on there. As a matter of fact, there's going to be *more* farming that we're going to do around that site than there is now." Hay will be grown on sections of farmland not occupied by the solar power plant, grassy berms will screen the view of the fenced-in solar arrays from the road, and sheep may even be allowed to graze among the panels. In all, the solar arrays will sit on less than one-third of the Kirby property, with large stretches of woodlands and wetlands left undeveloped.[2]

KDC's assurances have done little to lessen the ire of Jill Barone and a few dozen other activists who have rallied under the banner of a group called Stop Somerset Hills Power Plant. When I drove to meet Jill at her home, it was easy to find her neighborhood. All I had to do was follow a trail of roadside placards warning that the solar development trio—KDC, the Kirby family, and the pharmaceutical company Sanofi—would destroy Bedminster. Ironically, one of the rallying cries of this poster campaign was to "Save the Barn," a picturesque but tumbledown building on the Kirby property that had already been dismantled several months earlier, after being declared structurally unsafe by the town.

To fill me in on the reasons KDC's project would be so bad for Bedminster, Jill gathered several of her fellow activists for a casual supper around her and her partner Joe Iuoria's dinner table. Some lapsed into sentimentality about a time that, to all appearances, was long past. "We used to help Mickey Stevenson feed his pigs and milk his cows," one veteran said of another Bedminster farm decades ago. Turning to the present, he referred to the popular slogan now being used to promote Jersey-grown produce. "They call it 'Jersey Fresh,'" he said. "If they put solar panels on New Jersey's farmland, it'll be called Jersey Fresh-Baked." Others were candid about farming's limited role in Bedminster today; it's commonly used by large landowners as a shelter to avoid paying full residential property tax rates.[3] But everyone around the table agreed that fields of solar panels would be a blight on the landscape. Jill said the stand of trees separating her property from the solar facility won't sufficiently shield her view of the panels, and she also fears that lighting installed at the power plant will be an annoyance. (KDC claims there will be no outdoor lighting at the facility.) Others expressed worries about chemicals leaching into local groundwater from the panels and their galvanized steel mounts. And the six-foot-high perimeter fencing was another contentious topic. Even though KDC has proposed using less visible box wire rather than conventional chain-link fencing, some said it would still look ugly and would block the passage of wildlife through the property.

Former mayor Joe Metelski presided informally over the dinner table discussion. Opening a hefty folder of documents, he pulled out overlay maps of the site showing wetlands, soils, and critical wildlife habitat. He referred to Bedminster's recently adopted solar ordinance, which demands "special attention to protection of the rural character of Bedminster's countryside."[4] Then he read excerpts from Bedminster's master plan, which calls for protecting scenic vistas and preserving local farmland.[5] "Even liberals support this kind of stuff," he said to me in good-natured ribbing. My ties to "liberal" Massachusetts had come up earlier in the discussion, along with the implicit assumption that liberals are more prone to support solar energy than conservatives. (Jill's friends didn't seem to be aware of the pivotal pro-solar role that Barry Goldwater Jr. and other leading conservatives have played in Arizona and several other states.)

Just as it was easy for Jill and her friends to marginalize my views as those of a Massachusetts liberal, it was hard for me to take some of their concerns seriously. The Kirby farm was *not* a working farm, and hadn't been for some time. Its ten-acre residential zoning meant that, sooner or later, oversized houses with expansive, chemically treated lawns were the most likely alternative to KDC's solar plant. As Joe Metelski acknowledged, the owners of these suburban estates would be entitled to surround their properties with fences every bit as forbidding to wildlife passage as the fencing proposed by KDC. And would any sheep be found grazing on the manicured grounds of these multimillion-dollar manses? I had my doubts.

I was even more disappointed by the readiness of several of Jill's dinner guests to dismiss global warming as a serious threat. "We don't change the atmosphere," Jill's father Tom Metz stated flatly. He had recently moved back to New Jersey after several years in California. "It happens on its own." Then, backtracking slightly, he said that the real culprits in polluting the world are the Chinese and the Indians. "If everybody, right now, was doing what America's doing, we'd be just fine." Tom didn't seem concerned, or aware, that Americans continue to lead the global pack in per capita carbon emissions—ahead of China, India, and almost all other nations. Joe Metelski's perspective was no more reassuring: "Everything's cyclical, including the environment. Just hang around long enough and it'll go into another cycle."

Though I was troubled by some of the views I heard that evening, they do reflect the resistance that solar developers sometimes face when they propose to build large-scale projects on "greenfields"—natural areas or lands where farming, ranching, and forestry have been the primary prior uses. The specific issues may vary from the Northeast's fragmented farmlands and forests to the expansive deserts of the Southwest, but the tradeoffs in both contexts are real, and it's no surprise that the perceived sacrifices are generally felt most strongly by those living closest to proposed projects. Bedminster's current mayor, Steve Parker, put it diplomatically when we spoke shortly after my meeting with Jill and her friends. "Do you want an open space, agricultural, pastoral landscape, or do you want to achieve renewable energy? They both are good things, but at some point in time there are going to be decisions, and you are going to get one or the other."[6]

Bob Marshall, the assistant commissioner for sustainability and green energy at New Jersey's Department of Environmental Protection, has grappled with the same tensions. "We've spent over $1.5 billion to preserve farmland in the state," he told me. Letting these same lands get converted to large-scale solar power plants undermines the monetary and environmental value of this investment, in his view.[7]

The fate of the KDC-Sanofi power plant rests on decisions yet to be made by Bedminster's land use board and township committee. Month after month, the public deliberations drag on, with hundreds of pages of hearing transcripts dutifully prepared.[8] I sat through one of those meetings, running late into the night, and witnessed the hot tempers flaring as issues were hashed and rehashed. For both sides, it has become a war of attrition.

—◊—

Building solar projects on greenfields is tricky under any circumstance, but in the nation's most densely populated state, it verges on explosive. It's no surprise, then, that New Jersey is working so hard to come up with new solar sites on brownfields—potentially contaminated commercial and industrial properties that are vacant or underutilized. Developers of these sites enjoy special liability protection under the New Jersey Solar Act of 2012, which also requires the state to cover the incremental cost of installing solar projects on brownfields.[9] Acting on this mandate, the Board of Public Utilities authorized PSE&G, the largest utility in the state, to spend $247 million on forty-five new solar power plants. Of these, all but three are to be built on landfills and industrial sites.[10]

Even before the utilities board gave the green light to this new infusion of ratepayer funding in May 2013, PSE&G had completed an earlier round of five brownfield solar projects. Four are at gritty, disused properties it already owned: manufacturing plants for the coal-derived gas that fueled New Jersey streetlights a century ago. The fifth is a 13-acre solar array at the Kearny landfill, a small part of an unsightly stretch of solid waste dumps owned by the state and run by the New Jersey Meadowlands Commission. These giant mounds are all too familiar to anyone driving on the New Jersey Turnpike north of Newark or riding an Amtrak train just before it dives beneath the

Hudson on its way to New York's Penn Station. In my own travels, I've always looked at the Meadowlands as the very definition of "wasteland." I was heartened to learn that at least one small part of this despoiled area has been salvaged by the sun.

I met PSE&G's solar team leaders at the Kearny Landfill Solar Farm on a half-cloudy midsummer day. Though only 40 miles separated us from the Kirby farm in Bedminster, the distance felt much greater as I looked east, past several undulating rows of solar panels toward the lower Manhattan skyline. Turning to the south, I gazed at a rusting tangle of railroad bridges and highway interchanges, and beyond them, at Newark's stunted cityscape. Paul Morrison, project manager of the utility's Solar4All program, pointed to the gravel beneath our feet. "We're standing on 1,400 truckloads of dense-grade aggregate," he said.[11] It took close to two months to create a relatively even surface on this capped mountain of garbage, and about four months to install the 1,200 panels, anchored to heavy concrete footings like the solar arrays that Phil Beaumont is installing on contaminated sites in New Bedford, Massachusetts. The Kearny power plant's cost totaled $18 million, with $8.5 million coming from a federal grant under the American Recovery and Reinvestment Act.[12]

Todd Hranicka, PSE&G's director of solar energy, has learned a few lessons from all the heavy lifting at Kearny. As he surveys additional sites among the eight-hundred-plus landfills in PSE&G's service territory, he makes it a priority to minimize the need for major earthworks. At one point in the process, Todd's wife admitted to being bewildered by her husband's work. "You're spending a lot of time on landfills, and you leave [the house] in your suit and tie," she said to him. "Why do you dress like that? It's really got to be smelly and dirty." He explained: "Hon, these are like fields. They aren't active landfills." She still didn't get it: "So you're just walking on trash?" Once he showed her some photos, she understood. "The good sites that we have, they're just flat. They look like airport runways," he explained.

Once Todd finds a promising landfill site, he works on persuading local political leaders that a solar power plant would be a good fit. "Ten years ago it was the rage to build golf courses on landfills, but now they're all done," he explains. "There's no more golf needed in New Jersey." The public also has a limited appetite for converting land-fills to parks, he says. "People don't want their kids walking around on landfills." Solar power plants are a great alternative; they're a safe way to turn wastelands into productive assets.

The Kearny Landfill Solar Farm may have been costly to build, but it has one key advantage over some of PSE&G's other brownfield solar sites: elevation. This became painfully clear when Superstorm Sandy slammed into the Garden State in October 2012. The Kearny plant came away unscathed, but that was far from the case at the Lin-den PV array, erected at one of PSE&G's former gas plants just south of Elizabeth. "We built everything based on a hundred-year storm. This was a five-hundred-year storm," Todd recalls. The Linden array came online in May 2012, and five months later it was under seven feet of water. "In Solar 101, they don't teach you about your panels being under saltwater for twenty-four hours," he recalls. Electrical circuits were fried, panels were torn from their mounts, and an empty five-hundred-gallon fuel tank floated onto the site and landed right on top of a string of panels. Even though repairs began the day after the storm, it took a full six months to get the power plant running again. In an era of growing climate instability, Todd has a new criterion for siting solar projects: all major electrical connection points must be above the five-hundred-year floodplain.[13]

—⁂—

While closed landfills have emerged as popular sites for solar power, enterprising developers are also searching deep within the fabric of our older cities for new solar prospects. Chicago's West Pullman neigh-borhood hosts one of these installations, on heavily polluted acreage once owned by International Harvester. Exelon—a utility best known for its large nuclear reactor fleet—has built the solar plant, generating enough electricity for about 1,500 Chicago area homes. At the time Exelon City Solar went into service in March 2010, it was the largest inner-city solar power plant in America.

West Pullman was first settled in the late 1800s as an exclusively white working-class community serving factories developed by the West Pullman Land Association. For many decades, International Harvester (now Navistar) manufactured farm equipment there, alongside paint producers, metal casting plants, and other heavy industries that polluted the air and discharged a deadly brew of toxic chemicals into cisterns and storage tanks on their properties. Most of those plants are long gone, and the neighborhood—now over 90 percent African American—is riddled with gang violence and crippled by unemployment.

Carrie Austin is the alderman of Chicago's 34th Ward, a section of the far South Side that includes West Pullman. Reelected four times since 1994, she has fought hard to get corporations like Navistar to clean up the wastes they left behind when they fled the community. She has few kind words to say about the way Navistar cared for its property. "The previous owner of that site had no respect for our community at all. They just left it desolate, they left it contaminated, and they left it that way for over twenty years." Eventually the city placed a lien on the property, tore down the abandoned factory buildings, and began the long, costly process of remediating asbestos and other contaminants found at the site. By the time Exelon became interested in the property, the city had already spent $800,000 on this cleanup effort, but there was still a lot of work to do.[14] When Austin learned about Exelon's proposal to build a solar plant there, she was euphoric. "We'll take it!" she told the utility's exploratory team.

Part of what excited Austin was the prospect of converting a source of local shame into a place of hope and dignity. She imagined politicians, city planners, and solar developers from around the world coming to her forgotten corner of the city to witness this restorative reuse of a polluted industrial site. At the same time, she knew that people in her community would be skeptical about another outside company coming into their midst. A number of community meetings were held, and people wanted to know how the project was going to affect their household budgets as well as the community as a whole. At first people assumed that the project would actually lower electric rates in West Pullman, but that hope was quickly quashed. The project, at 10 megawatts, would feed its electrons into Exelon's regional power portfolio, which is dominated by nuclear and gas-fired power plants that are tens, and in some cases hundreds, of times larger in scale. There would be no discount for local customers.

The US Department of Energy's decision not to support the solar plant through a grant or loan guarantee came as an added disappointment—to Exelon as well as the West Pullman community. To Alderman Austin's relief, Exelon decided to move ahead with the project regardless. The $60 million facility would still qualify for a 30 percent federal investment tax credit, substantially reducing its ultimate cost to the company.

In July 2009 a minority-owned civil engineering firm, Environmental Design International (EDI), took charge of completing the pollution remediation, installing proper drainage, grading the site, and ensuring worker safety during construction. The goal was to have the facility up and running by early 2010, so the pressure was intense. Remediation work was still under way on one section of the property while workers were already driving solar mounting poles 12 feet into the ground in other areas. Not surprisingly, there were some unpleasant moments. One close observer recalls: "Every once in a while, one of these posts would just disappear." Construction would then be halted so that remaining cavities in underground cisterns could be filled and previously undetected storage tanks could be unearthed and removed. The visuals at these times were spooky, with workers in Tyvek suits and respirators wandering the site.

Deborah Sawyer, EDI's founder and CEO, describes the lower level of cleanup required for Exelon's solar installation as compared to other possible uses of this polluted parcel. "You don't have to clean the site to the point where somebody can eat off the ground if it's going to have a semi-industrial use," she says. You just have to make sure that any remaining hazards are stabilized so that no contamination will run off the site or get into the air. She offers a few rounded cost figures for remediating a site such as this one: "You can clean it up to 80 percent, let's say, for a million bucks. To go from 80 to 90 percent

Exelon City Solar brownfield installation in Chicago's West
Pullman neighborhood. (Photo credit: Turner Construction
Company/McShane Fleming Studios)

is $10 million. Then to go from 90 to 95 percent is $100 million."[15]
For Exelon City Solar, the least expensive of the three options proved
sufficient.[16]

Today, the clean geometrics of Exelon City Solar's PV arrays leave
no visible signs of the mayhem that scarred these grounds' acreage until so very recently. People living in homes bordering the property can
look out their windows at a neatly fenced sea of solar panels instead
of a forbidding industrial wasteland. There have been economic gains
as well, though not all of them felt directly by West Pullman residents. The property, leased by the city to Exelon, went back on the tax
rolls after many years of dormancy. Some two hundred people were
employed during construction of the plant, and while most of those
jobs went to tradespeople from outside the neighborhood, Alderman
Austin is happy that several of the apprentices hired from West Pullman ended up landing ongoing work in the solar industry. In addition,
a local welding shop enjoyed a temporary boost to its business when
it received the contract to produce the steel mounting posts—more

than seven thousand of them—that now hold up thirty-two thousand PV panels.

Despite these benefits, West Pullman remains a deeply troubled neighborhood. Stray bullets falling from the sky have shattered a few dozen solar panels so far—just one sad reflection of the ongoing street violence in this tough Chicago community. "No matter what it is I bring to the ward," Austin laments, "the violence still keeps us down. If you don't have the violence under control, who's going to move here?"[17]

Howard Learner, head of the Chicago-based Environmental Law and Policy Center, views Exelon City Solar more hopefully, as the bellwether of a broader solar transformation of degraded and disused industrial properties throughout the Chicago metropolitan area. Through a project called Brownfields to Brightfields, or B2B, his nonprofit advocacy group is scouring the area's industrial landscape to identify appropriate sites for solar development. So far, some two hundred vacant properties have been prioritized based on several criteria: their size (large enough to accommodate at least one megawatt of solar power); a history of contamination that makes them undesirable for other uses; the absence of shading by buildings or trees; and proximity to transmission lines, or to large commercial or industrial operations that can make direct use of the power.[18]

A drive with Howard through the far South Side left me impressed, and depressed, by the scale of some of the B2B prospects that his group has found. One of them is a 104-acre parcel occupied for nearly a century by the Acme Coke Plant. Giant ovens at this plant converted trainloads of coal into coke, a primary fuel source for the steel-producing blast furnaces that once lined the nearby Calumet River. After Acme shut its doors in 2001, there was some discussion about turning the site into a museum about the history of steelmaking, but little has come of the idea.[19] Most of the plant's buildings have since been demolished, but two tall smokestacks and a forlorn coal silo still stand as reminders of this filthy but once-bustling industry.

While some may have trouble looking beyond the industrial blight on brownfield sites like Acme, Howard sees an immense, largely dormant renewable energy resource. "Whether it's Chicago or Detroit or

Cleveland or Milwaukee," he says, "there are tremendous opportunities to unlock latent value in abandoned industrial brownfields and transform them into clean energy brightfields for the future." Realizing this potential, Howard knows, will depend on the building of a pragmatic alliance between policy makers and entrepreneurs. To get a few flagship projects going, his B2B team has begun to open up new lines of communications among government officials, solar companies, real estate developers, and civic groups in communities where some of the more promising sites are located. Exelon City Solar offers hope that, with shared determination, the B2B vision can be more than a redemptive dream.[20]

—m—

Just as closed landfills and abandoned industrial sites are part of our renewable energy frontier, some companies are scouting out solar opportunities on brownfields that they still occupy. Aerojet Rocketdyne is one of them. This firm has manufactured rocket propulsion systems for everything from intercontinental ballistic missiles to NASA space vehicles, and most of its manufacturing and testing has taken place on a 5,900-acre plot of land east of Sacramento, California. Aerojet's workforce peaked at twenty-three thousand during the 1960s, when the Cold War arms race and US-Soviet space rivalry were at their height. Today the company has three thousand employees spread across thirteen states, but it remains one of the leading rocket propulsion firms in America.

Aerojet may have carried America into space, but its activities here on earth created a serious enough health hazard to earn it a place on the US EPA's National Priorities List, the Superfund program's roster of America's most contaminated industrial sites. One of the culprits was trichloroethylene (TCE), widely used as a cleaning agent during manufacturing. Other prime pollutants were perchlorate, an oxidizer used in solid rocket fuel, and nitrosodimethylamine (NDMA), a component of liquid rocket fuel. Once traces of these chemicals began showing up in local drinking water, the polluted wells had to be closed. By 2013 Aerojet had pumped and treated 50 billion gallons of contaminated water at a cost of hundreds of millions of dollars. The water treatment effort is ongoing, but the removal of toxic soils has barely begun.

Michael Girard, Aerojet's sustainability director, meets me in the visitors' reception area. Bouncy and jovial, he invites me to ride with him to the 6-megawatt solar installation that now generates about one-fifth of the power needed to run his company's groundwater remediation program. The son of a fighter pilot, he makes sure I understand that he once was a "tree hugger" who attended college at Humboldt State, surrounded by California redwoods. He joined Aerojet in 1985, right after graduating with a degree in business. "Coming to Aerojet, I literally had to cut my hair," he recalls with a hearty laugh. Though his friends at first chided him for selling out, he takes pride in the work he has done to make Aerojet a cleaner, more environmentally minded company.

We drive past a few clusters of large, low-lying buildings where rockets of various shapes and sizes are produced. Michael tells me that drains carrying chemical residues from these factories have caused much of the pollution that still plagues Aerojet. Runoff from rocket test sites is the other big source of contamination. For many years, test site technicians did nothing to contain or treat the water they used to quench rocket flames. Rather, the polluted water spilled into nearby gullies and quickly migrated down into the underlying aquifer. The protective clay layer that might have blocked its flow had been destroyed decades earlier by gold mining dredges that had clawed their way deep into the earth, leaving long rocky mounds of spoils in their wake. Careless avarice had ravaged this landscape long before Aerojet came onto the scene.

As Michael and I continue our tour, the surroundings become almost pastoral. Cottonwood trees grow between mounds of dredge spoils, now half-concealed beneath a blanket of dry grasses. Here and there, golden poppies splash bits of bright orange on the machine-made slopes. Michael points excitedly to a few wild turkeys standing by the road. Then he spots a California jackrabbit. "You can tell the jack by the large ears and long legs," he explains. Coyotes, bobcats, and deer aplenty inhabit this eerie acreage. Though Michael has yet to encounter a mountain lion, he has seen its tracks.

"We're now entering one of our liquid test areas," my guide tells me as thick, buttressed walls of concrete come into view. The thrust of ignited rockets, strapped firmly to horizontal beds, is deflected upward by these bulwarks. Prior to tests, sirens sound and lights flash, warning people to clear the area. I imagine how utterly the tranquility of this large, fenced-in nature reserve must change at those fearsome moments.

A few hundred yards farther along, I see the first set of solar arrays, with their long rows of panels tilting slightly eastward to capture the late-morning sun. One cluster of solar fields is here; another is a short distance away. In all, there are twenty rectangular fields covering about 40 acres. The open space between them, Michael says, leaves room for the extraction wells and pump houses that operate in the area. The surface soil here is not contaminated, he reassures me, but the underlying groundwater is.

Michael parks alongside one of the solar arrays, but he asks me to wait by the car. He wants to make sure no rattlesnakes are resting in the shade beneath the plant's electrical equipment. He finds no snakes, but we do see several wild turkeys cooling themselves under some nearby solar panels.

It was on this very spot that Arnold Schwarzenegger chose to commemorate one of California's biggest milestones in the battle against climate change. On September 15, 2009, in blazing sunlight with thousands of PV panels spread out behind him, the actor-turned-governor signed his now-famous executive order calling on the California Air Resources Board to ensure that 33 percent of the state's retail electric power would come from renewable energy resources by 2020.[21] "I can't tell you how excited I was," Michael recalls. "It was a big deal for us to be suddenly held in a positive light relative to our Superfund status."

Getting a solar farm built on a Superfund site was no small feat, but Michael was determined to make it happen. Queries were sent to a number of solar companies, and one developer called SPI (Solar

Power Incorporated) came in with the best combination of technical know-how and financial savvy. Another key player was the Sacramento Municipal Utility District (SMUD), a public utility company serving Aerojet and the surrounding area. A longstanding leader in promoting renewable energy, SMUD offered SPI a package of rebates totaling $13 million. With those rebates, the federal investment tax credit, and land donated by Aerojet, a project nominally costing $20 million ended up requiring just a few million dollars in capitalization. In the end, Aerojet was able to sign a deal with SPI securing the solar plant's power for less than the rate it was paying SMUD for the rest of its electricity.

Looking around me as I stand in this remote corner of the Aerojet campus, I ask Michael whether the company might build more solar power any time soon. After all, today's installation covers little more than half a percent of the company's land. He tells me this is under consideration. On one hand, the SMUD rebates have shrunk, and though he doesn't say so explicitly, my impression is that Aerojet would not garner much of an added publicity punch from expanding its solar capacity. On the other hand, PV prices will very likely continue to drop, and solar power is an excellent reuse of property that the EPA gingerly calls "otherwise encumbered."[22]

Michael has clearly found ways to make environmental performance a central facet of corporate life at Aerojet. He believes in taking action, even if technological change may soon render today's investments outmoded. "It's always a learning process," he says. "Twenty or thirty years from now, this will be antique." He gazes at the solar arrays that he has so successfully planted on this desecrated soil. "You've got to do something today because, if not, you basically continue to wait for tomorrow and then nothing gets done."[23]

—ᴍ—

America's brownfields could be viewed as a pervasive liability or a momentous opportunity. Taking the latter view, the EPA has screened over sixty-six thousand brownfield sites to gauge their renewable energy potential. The surveyed sites cover more than 35 million acres and range from mines and factories to hazardous waste dumps and solid waste landfills. While the agency's RE-Powering America's Land Initiative has also assessed the wind, biomass, and geothermal resources that could be harvested on these lands, its findings regarding solar are quite astounding.

Overall, the EPA's mappers have identified a total of 5,500 giga-watts (5.5 trillion watts) of solar potential on the brownfield sites it surveyed. This translates into enough electricity for about 800 million American households, or seven times the power consumed by our nation's 116 million households.[24] The agency is careful to point out that it has not conducted detailed technical and economic analyses specific to most of these sites, but the overall picture is encouraging. A partial EPA listing of now-operating brownfield solar projects includes landfills in twelve states, Superfund and other federally designated hazardous waste sites in ten states, and abandoned manufacturing plants, mines, and munitions plants in several others.[25] A number of these installations are on US government property, and the EPA has identified many additional prospects at military bases, research labs, and other federal facilities. So far, 320 government properties have been screened favorably for sizeable PV installations.[26] This has been a key step toward fulfilling President Obama's call, in December 2013, for a stepped-up federal investment in renewable energy projects on government-owned brownfields.[27]

Along with benefiting the environment, brownfield PV projects have to be economically viable. A lot depends on local electric rates: in markets with ready access to natural gas and in areas that continue to rely on cheap coal, it's hard for renewable energy projects to compete, especially if extra measures are needed to remediate pollution at brownfield sites. Yet these added site preparation costs are often offset by lower lease payments on polluted properties that have little market value. Moreover, there are sources of support specific to brownfields that these energy projects may be able to draw upon. An EPA grant for assessing and cleaning up a contaminated site is one option. A loan from one of the EPA's brownfield revolving loan funds is another.[28]

Adam Klinger is a policy analyst who has worked on a variety of EPA programs over the past twenty years. He has written clean air and hazardous waste cleanup regulations, and has contributed to the EPA's smart growth and green building programs. Now he devotes his efforts to broadening public awareness of the clean energy opportunities that lie hidden behind the chain-link fences of landfills and industrial sites. When states, counties, or municipalities have brownfield sites suitable for renewable energy projects, he lets them know. When developers get nervous about the liability that they may incur if they build or operate a renewable energy project on contaminated property, he makes sure they receive accurate information on the protections they will enjoy so long as they don't aggravate the existing pollution hazards. (EPA's enforcement office prepares what he calls "comfort letters" addressing these concerns.) And when new brownfield projects come on line, he works with EPA's media team to bring these success stories to light.

For Adam, reviving brownfields is much more than a matter of policy; it's a personal mission. "You walk by places and properties that aren't living up to their potential and aren't contributing to the community, whether it's because they're blighted or because they're contaminated or because they're under-used. . . . There's just a vibrancy that the community is missing." Bringing these properties back into productive use is what gets him charged up about his job. When Harvard recognized the RE-Powering America's Land Initiative as one of the Top 25 Innovations in American Government in 2013, he was thrilled. "We're excited that people are . . . intrigued and want to join the movement," he tells me.[29]

EPA's mapping of brownfield sites invites us to fathom the magnitude of the clean energy resource that lies within our reach, on lands that too often are shunned as irredeemable wastelands. If solar power plants are built on even a fraction of this real estate, America's brownfields can become vital links to a lower-carbon future.

CHAPTER FIVE

The Desert's Harvest

TO SUPPLY 100 PERCENT of America's power needs from the sun, the National Renewable Energy Laboratory (NREL) has estimated that solar installations would occupy about 0.6 percent of the nation's total land area. That's less than 2 percent of the land now in US crop production, but it's still a huge stretch of terrain: about 21,000 square miles—a little less than the size of West Virginia and about twice as big as my home state, Massachusetts.[1] Solar technology's hunger for land highlights why we shouldn't expect solar arrays on buildings, parking lots, and brownfields to satisfy all our electricity needs.

Rooftop solar arrays can meet most of the home power demands of millions of American families, but millions of others don't have adequate roof access or solar exposure to produce their own electricity.[2] Brownfields offer great promise as future solar venues, but identifying those sites, resolving competing land uses, and making them safe for solar development can take a lot of time and cost a great deal of money. And what about all those high-rise office buildings, factories, and transportation systems that consume much more electricity than they can possibly supply through on-site solar harvesting? NREL estimates that it would take 181 square meters (1,948 square feet) of solar surface area to meet the average American's share of our total power consumption. That's about one-fifth of the "urban area footprint," or land area available per capita, in our towns and cities.[3]

Solar power opportunities within the fabric of our built environment may face a number of constraints, but America's overall land resources are a solar power cornucopia, as the advocacy group Environment America has documented in its recent report, *Star Power: The Growing Role of Solar Energy in America*. The authors of that report start with state-by-state estimates of solar power's technical potential as reported by the NREL. In gauging that potential, NREL has taken both photovoltaics and concentrating solar power (CSP)[4] into account, and has been careful to exclude land areas deemed unsuitable for solar development. Designated wilderness, nature conservation, and public recreation areas are sidelined along with wetlands and historic sites. Land with a slope greater than 3 percent is also rejected, given the difficulty of building ground-mounted solar installations on uneven terrain.[5] Environment America then compares NREL's state-by-state projections with current electricity use in each state, and its findings are staggering. From a low of two times current electricity use in smaller, heavily settled states like Massachusetts and Rhode Island, solar's potential rises to 1,433 times New Mexico's total power consumption.[6] (See fig. 5.)

In a country with such vast solar resources, it's not surprising that solar developers are busy building giant solar power plants outside the fabric of our built environment. Predictably, many are investing their greatest hopes in the Southwest, where long sunny days and vast stretches of open desert offer seemingly ideal conditions. These developers typically deploy hundreds of thousands, and sometimes millions of PV panels in solar fields that can stretch across a few thousand acres. When the sun is shining, these installations can create electricity on a scale equivalent to many of our conventional power plants. Generally cheaper to build than smaller solar arrays, utility-scale solar plants have grown from being minor players in the solar industry as recently as 2010, to accounting for more than two-thirds of newly installed solar capacity today.[7] According to industry analysts, the predominance of these larger solar facilities is expected to continue during the coming years. (See fig. 6.)

I first encountered utility-scale solar power in the Eldorado Valley, about 20 miles east of Las Vegas. To get there from McCarran International Airport, I drove along the Great Basin Highway through unrelenting sprawl until, just beyond the Railroad Pass Hotel and Casino in the town of Henderson, the landscape opened onto a broad expanse of desert. The contrast was stark, sudden, and welcome.

Fig. 5. *Solar Electricity Potential by State (in Gigawatt-Hours)*

STATE	POTENTIAL SOLAR ELECTRICITY CAPACITY (GWH)	TOTAL 2012 ANNUAL RETAIL ELECTRICITY SALES (GWH)	NUMBER OF TIMES SOLAR ENERGY COULD POWER THE STATE
Alabama	3,758,165	86,183	44
Alaska	8,283,142	6,416	1,291
Arizona	24,556,070	75,000	327
Arkansas	5,023,834	46,860	107
California	17,699,253	259,538	68
Colorado	19,452,241	53,685	362
Connecticut	33,961	29,492	1
Delaware	289,375	11,519	25
DC	2,499	11,259	0
Florida	5,274,479	220,674	24
Georgia	5,566,467	130,979	42
Hawaii	57,127	9,639	6
Idaho	7,466,971	23,712	315
Illinois	8,224,624	143,540	57
Indiana	4,992,152	105,173	47
Iowa	7,029,897	45,709	154
Kansas	22,515,073	40,293	559
Kentucky	1,862,803	89,048	21
Louisiana	4,184,643	84,731	49
Maine	1,105,986	11,561	96
Maryland	629,350	61,814	10
Massachusetts	111,397	55,313	2
Michigan	5,290,013	104,818	50
Minnesota	10,840,506	67,989	159
Mississippi	5,016,233	48,388	104
Missouri	5,381,978	82,435	65
Montana	9,741,194	13,863	703
Nebraska	14,131,977	30,828	458

Fig. 5 (*continued*)

STATE	POTENTIAL SOLAR ELECTRICITY CAPACITY (GWH)	TOTAL 2012 ANNUAL RETAIL ELECTRICITY SALES (GWH)	NUMBER OF TIMES SOLAR ENERGY COULD POWER THE STATE
Nevada	16,945,868	35,180	482
New Hampshire	63,453	10,870	6
New Jersey	499,848	75,053	7
New Mexico	33,208,762	23,179	1,433
New York	1,574,149	143,163	11
North Carolina	4,329,556	128,085	34
North Dakota	9,777,286	14,717	664
Ohio	3,742,742	152,457	25
Oklahoma	14,472,440	59,341	244
Oregon	6,586,711	46,689	141
Pennsylvania	631,733	144,710	4
Rhode Island	17,135	7,708	2
South Carolina	2,803,221	77,781	36
South Dakota	11,645,189	11,734	992
Tennessee	2,295,918	96,381	24
Texas	62,153,732	365,104	170
Utah	10,290,431	29,723	346
Vermont	57,475	5,511	10
Virginia	1,932,186	107,795	18
Washington	1,947,153	92,336	21
West Virginia	59,938	30,817	2
Wisconsin	5,111,137	68,820	74
Wyoming	11,142,414	16,971	657

Sources: Judee Burr, Lindsey Hallock, and Rob Sargent, *Star Power: The Growing Role of Solar Energy in America*, Environment America Research and Policy Center, table B-1, November 2014, http://www .environmentamerica.org/. Solar production estimates include photovoltaic and concentrating solar power, and are based on data from NREL, *U.S. Renewable Energy Technical Potentials: A GIS-Based Analysis*, July 2012, http://www.nrel.org/. Protected wilderness, nature conservation, and recreation areas, wetlands, historic sites, urban areas, and land with greater than a 3 percent slope are excluded. Data on annual retail electricity sales by state are from the US Energy Information Administration.

Fig. 6. *Annual Installed Photovoltaic Capacity by Sector (2009–2016)*

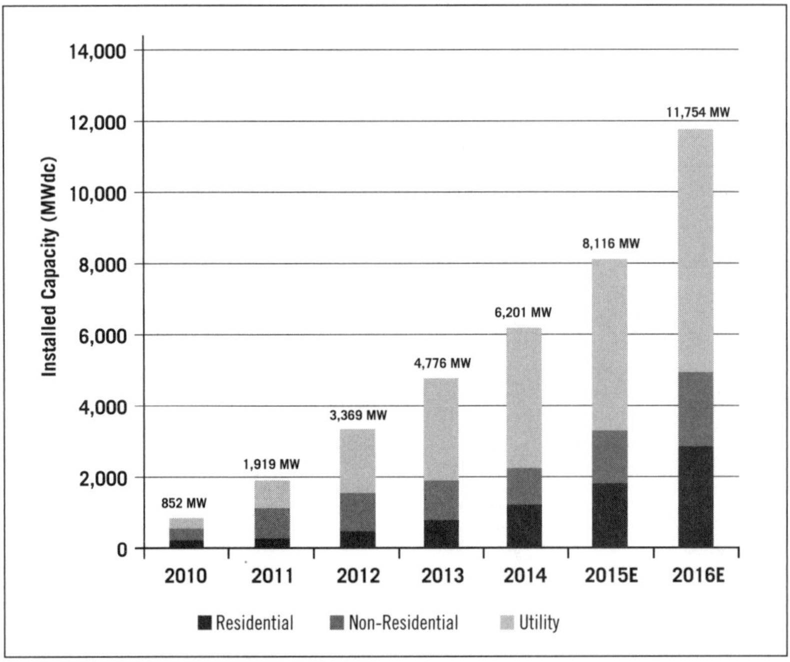

Source: SEIA/GTM Research, *U.S. Solar Market Insight Report—2014 Year in Review,* http://www .greentechmedia.com/research/ussmi. Megawatts are measured in direct (DC) current, prior to conversion to alternating (AC) current.

The Eldorado Valley stretches for miles, but much of it lies within the municipal boundaries of Boulder City, a quiet community of 16,000 on the edge of Lake Mead, whose leaders have worked hard to keep metropolitan Las Vegas from expanding into their midst. As far back as 1931, when the federal government built Boulder City as a dormitory for construction workers at the Hoover Dam, the community's planners sought to insulate its inhabitants from Sin City, with its drinking, gambling, and organized crime. The town's ban on alcohol was eventually lifted, but Boulder City remains one of only two localities in Nevada to prohibit gambling.

During the 1980s and 1990s, Boulder City's margin of separation narrowed as new suburbs sprang up—products of the Las Vegas real estate boom. To create a physical buffer, the city struck a deal with the federal government in 1995, annexing 160 square miles of land in the Eldorado Valley with the promise to set most of it aside as a conservation easement protecting the desert tortoise and other native species. A few years later, local officials voted to designate a small fraction of this land as the Boulder City Solar Energy Zone. Today it is America's largest municipally owned solar energy zone, with more than 8,000 acres leased to half a dozen solar developers.

"This was a piece of dirt that I was getting nothing for," Boulder City mayor Roger Tobler tells me when we meet at City Hall to discuss the solar zone's rollout. "And now we've turned it into a major money-maker for the city." Tobler owns and runs a local hardware store when he's not performing his part-time duties as mayor. He smiles across the uncluttered surface of his City Hall desk, occupied by little more than a computer monitor and a Super Big Gulp soda. Lease payments from solar projects have salvaged the city's budget during tough economic times, he acknowledges. I notice the crest on his surf-green polo shirt, emblazoned with the words "Boulder City—Home of Hoover Dam," and am reminded that renewable energy in its various forms has been a key economic driver since the community's earliest days.

Mayor Tobler is sanguine about what solar power can bring to the desert lands within his city's borders. "What else is going to be out there? It's not going to be a strip mall. It's not going to be a Nordstrom's," he says. "It's going to be a great place to develop renewable energy." The leases with solar developers will provide his city with a continuous income stream for decades to come.[8]

One of the solar energy zone's principal developers is Sempra Energy, a Fortune 500 energy services holding company based in San Diego. Though most of Sempra's power generation assets are natural gas plants, its Copper Mountain solar projects in the Eldorado Valley are mainstays of the company's growing renewable energy portfolio. When the last of Copper Mountain's three phases is complete, enough solar power will be produced to meet the needs of 145,000 homes in

Southern California. That electricity will be sold to the Los Angeles Department of Water and Power, Pacific Gas & Electric, and the City of Burbank.[9]

California's renewable portfolio standard has created a major stimulus to utility-scale projects like Copper Mountain. Driven by the mandate that 33 percent of California's power must come from renewable energy by 2020, the state's utilities are scrambling to line up solar, wind, geothermal, and other qualifying power suppliers to fulfill their compliance quotas. The California Public Utilities Commission has given preference to in-state sources of renewable energy, but—luckily for Sempra and other solar developers in the Boulder area—a green light has been given to out-of-state suppliers that feed their power directly into the California grid. By running a six-mile power line to a transmission substation owned by the Southern California Public Power Authority, Sempra has the linkage it needs to qualify as an in-state resource.

An inspiration to community leaders across the Southwest who are eager to jump on the solar bandwagon, the Boulder City Solar Energy Zone has become a whistle-stop for national politicians who want to be seen as renewable energy leaders. In April 2011 Senate Majority Leader Harry Reid spoke glowingly about "all those hard hats at work" when he visited the construction site of Copper Mountain Solar 1 on the first day of his "Driving Nevada Forward" campaign tour.[10] (Construction of the project's three phases will employ more than 1,300 on-site workers.)[11] A year later, Barack Obama used a shimmering field of Copper Mountain PV panels as the backdrop for a rousing speech about solar energy's role in creating clean energy jobs and weaning America off foreign oil.[12]

Lisa Briggs remembers the escalating pressure as her public relations team at Sempra prepared for the president's appearance. On the first phone call, she was told: "Somebody from the administration may want to stop by for a photo." Two days later, there was another call: "Definitely someone from the administration is going to want to stop by for a photo." In another two days, she learned: "It's the president." By the time a nine-person advance team met her at the site, the "photo stop" had been elevated to a two-and-a-half-hour visit at which the president would make a major policy announcement. Trucks were brought in to block off rows of panels, and five hundred feet of banners were strung up along a border fence to obscure a potential shooter's line of sight to the podium where the president would speak. When the president finally came, F-16s were flying overhead. "It was a once-in-a-lifetime experience," Lisa says before adding with a laugh, "I really *hope* it was a once-in-a-lifetime experience!"[13]

Just a few months after Obama's visit to Copper Mountain, his administration released the final environmental impact statement for the federal government's Roadmap for Utility-Scale Solar Energy Development on Public Lands. This bold initiative lays out seventeen Solar Energy Zones in six southwestern states, covering 285,000 acres of federal lands. The program's goal, in the words of Deputy Secretary of the Interior David Hayes, is to create "an enduring, flexible blueprint for developing utility-scale solar projects in the right way, in the right places, on our public lands." According to official assessments, full development of solar energy resources on these lands may yield up to 23.7 gigawatts of solar energy, enough to meet the needs of seven million American homes.[14] Within the approved zones, solar developers are to be granted expedited permits by federal agencies. Beyond the zones, the government has identified another 19 million acres of federal land where "carefully sited" solar development can take place. Nearly 79 million acres have been sidelined as a poor match for solar development because of their natural and cultural resources.[15]

The vastness of the federal government's solar energy zone or "SEZ" designations points to a vexing aspect of utility-scale solar power: its hunger for land. Joe Rowling, Sempra's vice president for project development, addresses this issue as we walk down a dirt access road between two solar fields at Copper Mountain. Long rows of solar panels stretch toward the horizon; each row is carefully spaced to avoid any early-morning or late-afternoon shadows cast by adjacent panels in adjacent rows. By the time this project is fully built,

2.4 million solar panels will blanket more than 2,800 acres.[16] "These facilities have a huge footprint," Joe says. "And it's necessarily so, because they're collecting a very diffuse source of energy. If it were otherwise, we wouldn't be able to go outside without getting fricasseed!"[17]

—⁂—

To many solar proponents, the Southwest's desert areas are an unexploited treasure trove that should be put to productive use promoting a cleaner energy future. Mayor Tobler expressed a particularly blunt version of this utilitarian view when he likened the land now dedicated to the Boulder City Solar Energy Zone to "a piece of dirt." In his speech at Copper Mountain, President Obama reflected his own pragmatic perspective, heralding his administration's success in approving new solar projects on "public lands that aren't otherwise being utilized."[18] This basic premise has been vehemently challenged by a number of dedicated conservation groups.

One outspoken critic of the president's push for solar development on public lands is Janine Blaeloch, director of the Western Lands Project in Seattle. Blaeloch's organization has an unequivocal mission: "to keep public lands public."[19] In pursuing this goal, it gives no greater regard to renewable energy projects than to the timber companies, oil and gas drillers, and mining operations that have been profiting from public lands for more than a century. In February 2013 the Western Lands Project joined two other conservation groups in filing a lawsuit challenging the government's Solar Energy Zone designations. "The administration is opting to needlessly turn multiple-use public lands into permanent industrial zones," Blaeloch said when the suit was filed. "Solar development belongs on rooftops, parking lots, already-developed areas, and on degraded sites, not our public lands."[20]

While some conservation and recreation groups share Blaeloch's categorical rejection of solar energy projects on open lands, most mainstream environmental organizations like the Sierra Club, the Natural Resources Defense Council, the Center for Biological Diversity, and Defenders of Wildlife have taken a less doctrinaire approach.

They recognize the problems that can result from solar installations encroaching on natural areas, but they view those impacts in the broader context of the tough energy choices we face as a nation. Defenders of Wildlife's California program director Kim Delfino states it eloquently: "We strike a balance between addressing the near-term impact of industrial-scale solar development with the long-term impacts of climate change on our biological diversity, fish and wildlife habitat, and natural landscapes."[21] At times, in striking this balance, groups find themselves filing objections to specific proposals that they see as posing an elevated risk to natural resources. As often, though, they team up with solar companies in developing detailed plans to mitigate the adverse environmental impacts of large-scale projects. California Valley Solar Ranch, on the Carrizo Plain about halfway between San Luis Obispo and Bakersfield, is one such project.

The Carrizo Plain is sometimes generously called California's Serengeti, though it only covers about 400 square miles, less than one-tenth the size of that expansive East African savanna. Though parts of this semi-arid plain have been used for farming and grazing, Carrizo is still noted for its biodiversity—the tiny, long-eared San Joaquin kit fox, the giant kangaroo rat, burrowing owls, the reintroduced pronghorn antelope and Tule elk, and several other animals listed as endangered or threatened under state and federal law. The area is also blessed with more than three hundred days of sunshine per year and is relatively close to the major population centers of Southern California, making it one of the nation's best spots for producing solar power.

Solar developers were drawn to the Carrizo Plain as early as the 1980s, when ARCO Solar built one of the nation's first utility-scale photovoltaic fields in California Valley, an unincorporated community in the northern part of the plain. The plant used mirrors to concentrate the sun's energy on crystalline silicon solar cells in an attempt to maximize the output of a technology whose per-unit cost was still prohibitively high. Unfortunately, by magnifying the sun's intensity, the mirrors caused the PV cells to discolor and eventually lose power. In any case, the plant couldn't compete with cheaper conventional electricity. Nikki Nix recalls the bitter stories of people who built and then dismantled the project. Her father worked for a time as a maintenance man at the plant. "Right on the box, it said 'Not for mirror enhancement,'" she was told of the PV panels used at the site. Reminders of this short-lived venture are still around: the metal trusses that once braced the mirror assemblies now serve as fence rails and carport rafters at many local homesteads.

The roots of distrust and disillusionment among California Valley's residents predate ARCO Solar's debacle, going back to the community's hyped-up launch as a residential subdivision in the early '60s. Seven thousand 2.5-acre lots were put up for sale, but only two hundred families ended up building homes in the area. When Nikki graduated from eighth grade, she was valedictorian—one of two students in her grade. For high school, she had to board in a town 50 miles away.

When local people learned a few years ago that SunPower, a Silicon Valley–based company, was looking to build a solar plant in their midst, they voiced their misgivings. Would this be another hit-and-run fiasco like the ARCO Solar plant? Would their aquifer be sucked dry by the water needed to clean all those PV panels and keep road dust down at this new solar farm, dozens of times larger than the ARCO Solar plant? And then, when it became known that SunPower planned to set aside thousands of acres as conservation lands, some residents fretted that their recreational enjoyment of the area would be stifled. Would their grandkids still be able to ride their dirt bikes across all those absentee-owned properties that had become the community's extended backyard?

SunPower's development team decided that someone familiar with the local community's bumpy history and wary residents was needed. Nikki, who had been working as a real estate agent in the area, was an obvious choice. Hired as the company's community relations manager, she listened to people's concerns. "Whether you own two-and-a-half acres or twenty, you start to feel like all this open space is yours," she explains. "You get that all-in-my-view-is-mine kind of thought." As she laid out SunPower's plans at local meetings, she quickly realized that the "Yay! Green energy!" mantra would not win over the project's neighbors. Instead she focused on the hundreds of construction jobs and the $315 million that the project would pump into the area's economy over its quarter-century lifespan. "I spend a lot of time talking to folks," she says. "It seems to put their minds at ease."[22]

Early in the morning, I set out in a rented car from Bakersfield, home to the Kern River Oil Field, one of California's earliest oil discoveries and still one of its most prolific oil producers. A scarred moonscape dotted with thousands of oil derricks, just north of town, and the high school football team's moniker, the Drillers, stand as testaments to this fossil fuel legacy. By mid-morning, I have climbed out of the San Joaquin Valley and crossed the Temblor Mountains, part of the California Coast Ranges. As I descend onto the Carrizo Plain, it's easy to spot the California Valley Solar Ranch. Some large fields are already filled with orderly rows of solar arrays. Other fields have thousands of steel posts sticking out of the ground, with pallets of PV panels stacked nearby.

Nikki is waiting for me outside a cluster of simple white trailers, field offices for SunPower and the various subcontractors that were working on the project. She leads me to a small office in one of the trailers, where we are joined by Bill Alexander, the on-site representative of NRG Energy. This is the same company whose customer affairs chief, Tom Gros, guided me through the solar installation at Patriot Place. As typically happens, a developer like SunPower takes all the early steps to launch a power plant like this one, applying for the necessary permits, buying or leasing the land, meeting federal and state environmental protection requirements, and lining up a long-term power purchaser for the electricity generated by the facility. Then it sells a construction-ready project to a company with deeper pockets—in this case, NRG Energy. SunPower's ongoing role is vital, however. As the engineering, procurement, and construction or "EPC" contractor, it bears lead responsibility for designing the facility, preparing the site, buying and installing all the equipment, and delivering a ready-to-run power plant to NRG. It will also operate the solar farm in its early years.

During construction, a third firm—Bechtel—handles the day-today logistics, hiring and managing all the unionized tradespeople that are needed to build the plant. At peak times, there have been as many as six hundred workers on-site—carpenters, electricians, pile drivers, ironworkers, crane operators, and more. It's no small feat to install 750,000 panels along with the motor-driven, GPS-guided tracking systems that keep them aligned with the sun. Converting the DC current generated by the panels to 250 megawatts of grid-ready AC current is a further hurdle. And then the electricity's voltage has to be boosted so that the project's power purchaser, Pacific Gas & Electric, can deliver it economically to its customer service territories.

Bill Alexander maintains an on-site presence to make sure that all steps in this giant logistical puzzle are in line with his company's expectations. But his real hands-on work is managing a team of biologists who are tasked with ensuring the project's compliance with a formidable set of environmental safeguards. Some of these requirements are laid out in a tome prepared by San Luis Obispo County's board of supervisors. Others come from a settlement agreement reached with three national environmental groups: the Center for Biological Diversity, Defenders of Wildlife, and Sierra Club.[23] Bill's academic degrees—two in biology, one in forestry—serve him well as he tackles these tasks.

Before we head out to the solar fields, Bill invites me to share what he calls an "international potluck lunch" with nine biologists who are stationed at a cluster of computer monitors in one of the trailers. All are young women who commute from what Nikki calls "civilization"—towns near the coast, more than an hour away. (At least half of the project's workforce is required by the county to travel to and from the site in van pools, and another 25 percent must carpool. This reduces road accidents, minimizes risks to wildlife in the area, and helps shrink the project's carbon footprint.) Along with enjoying the lunchtime camaraderie, the gathered scientists clearly relish the eclectic array of Mexican and Asian dishes—much better than what's offered at a small taco truck that parks outside a neighboring ranch, referred to as the "roach coach."

After lunch, Bill, Nikki, and I hop into Bill's truck and head out for a tour of the project. We drive slowly—no more than 15 miles per hour—on an unpaved access road, passing a "water buffalo," a tanker that showers water on the decomposed granite road surfaces to keep dust levels down. In addition to reducing fugitive dust, sprinkling the roads may help reduce workers' exposure to the soil-borne fungus that causes Valley Fever. There's been a major spike in the incidence of this flu-like disease since construction began.[24] Near the main road, Bill slows to a near stop as he steers his truck over a set of steel grates where water pummels his vehicle's wheels and undercarriage. This bath, required on entering or exiting the site, is intended to keep public roadways clean and limit the spread of unwanted flora—the seeds of bromes and other invasive plants that generations of farmers and ranchers have brought into the area. Once the project is built, native grasses will be planted to restore at least some semblance of the native Carrizo habitat. Bill mentions that sheep grazing may even be introduced as a natural control on grass growth between rows of solar panels.

From a hilltop where power lines from the solar farm join PG&E's high-voltage grid, we get a great view of the solar fields stretching across the parched flatlands below. We speak briefly there with two archaeologists on SunPower's payroll; their job is to identify any Native American artifacts uncovered during construction. They show me a rock with a round depression, likely used as a mortar bowl by one of the tribes that once inhabited this landscape. Native American monitors, also hired by SunPower, decide which items are worth repatriating to their tribes.

Then we descend to a solar field where workers in hard hats are using pile drivers to pound twelve-foot steel poles halfway into the earth. These poles will soon cradle the rotating tubes to which long rows of sun-tracking panels are affixed. In another field, we witness the quick handiwork of several crews as they snap panel after panel into place. Bill calls these simple assemblies Tinker Toys. The trick, he says, is making sure the flow of hardware is brisk and well sequenced.

Before any site work could begin on the solar farm, teams of consultants carefully surveyed the area to identify local wildlife and develop plans for protecting particularly vulnerable species. One target was the San Joaquin kit fox, a diminutive creature no larger than a housecat that has lost most of its habitat to farms, oil and gas operations, and cattle raising in the San Joaquin Valley. While some kit fox have adapted by becoming urban scavengers in Bakersfield and other towns, the Carrizo Plain is one of the few areas where these animals have survived in nature, hunting for rodents by night and taking shelter from coyotes in their underground dens. Bill and his biologists have taken pains not to endanger the dozen or so kit fox living in and around the project. Following agreed-upon protocols, they have evacuated the animals from active dens in construction areas, fenced off buffer zones to keep the fox from returning during the construction period, and built artificial dens out of buried sections of culvert pipe in mitigation areas away from the solar fields. "NO kit fox was harmed in any way," he assures me emphatically.

The overall layout of the solar farm has also been adjusted to accommodate movement through the area by kit fox, pronghorn antelope, and Tule elk. Originally the plan had been to create very large blocks of solar arrays with little space between them. This, together with the acreage occupied by solar fields at the nearby Topaz Solar Farm, alarmed Ileene Anderson, a wildlife biologist at the Center for Biological Diversity, who knew that the pronghorn and elk needed to migrate unimpeded. Under pressure from her organization and two other environmental groups, SunPower redrew the map to create broader wildlife transit corridors between multiple, dispersed solar fields.[25] Around the individual solar arrays, SunPower committed to use special wire mesh with four-inch openings sufficiently large to let kit fox through. Surrounding the perimeter of the facility as a whole, it put up fencing with 18 inches of ground clearance—enough to let pronghorns pass uninjured beneath the lowest strand of smooth wire.

Prized prey for the San Joaquin kit fox, the giant kangaroo rat is another species that Bill's biologists have worked hard to protect. This endangered rodent with a leap reminiscent of Australian marsupials can only be found in about 2 percent of its historic habitat, and the Carrizo Plain is one of those remaining refuges. In a stretch of dry

Aerial view of California Valley Solar Ranch in San Luis Obispo County, California. (Photo credit: SunPower Corporation)

grassland between two solar arrays, Bill points to one of the "GKR condos" that his staff has built for giant kangaroo rats that were re-located from construction zones. In each of these, a small area just a few feet across is enclosed with fine wire mesh, keeping rats in and predators out. Once the animals have adapted to their new surround-ings by digging new burrows into the soil, the fences can be removed. Of 244 rats that were targeted for relocation, half a dozen died while being trapped—an acceptable loss under the terms of the "incidental take" permit issued by the California Department of Fish and Game. The survivors will be free to reinhabit the solar fields once construc-tion is complete.[26]

Burrowing owls, the California condor, the blunt-nosed leopard lizard, and a whole host of protected plant species also command the attention of Bill's biologists. Despite all these efforts, there is no denying that utility-scale solar farms transform local ecology by vir-tue of the land they occupy. Solar fields and related facilities at the California Valley Solar Ranch cover about 1,500 acres. The project's overall boundaries include another 3,200 acres, though much of this has been set aside for conservation. To soften the blow to the area's natural resources, the county obligated SunPower to dedicate thousands of additional acres for conservation. Ileene Anderson and her colleagues at Defenders of Wildlife and the Sierra Club negotiated yet a further increment of mitigation lands. In all, more than 12,000 acres in the Carrizo Plain will be protected through outright purchases and conservation easements—about eight times the facility's actual footprint.[27]

—∾—

By the time I leave the Carrizo Plain and head back to Bakersfield, it's late afternoon. Passing through the eastern foothills of the Temblor Range, I am struck by the barren ugliness of my surroundings. Oil derricks nod hypnotically on both sides of the road, their slow animation standing out against the stillness of the storage tanks, gas wells, and steel pipelines that are strewn across the terrain. Eventually this scarred land yields to endless irrigated fields and the stench of sprawling cattle feedlots—all icons of the San Joaquin Valley. I can't help thinking back to the kit fox, kangaroo rats, and other creatures that once had thrived in the area. Why were the oil prospectors and agribusinesses that have profited so handsomely from cheap mineral leases and subsidized water allocations never called to account for the havoc they have wreaked upon this ravaged landscape?

Conflicts between land conservation and development of the state's natural resources have run through California's history ever since John Muir fought successfully to create the Yosemite and Sequoia National Parks in 1890. These tensions only grew as the state's population multiplied, farms expanded, and extractive industries spread across the land. Over the past decade, the press to clean up California's energy economy has spawned some significant new stresses. With its call for 33 percent of electricity to come from qualifying renewable sources by 2020, the state's renewable portfolio standard (RPS) has created a robust market for new solar projects and wind farms, and regulations favoring in-state renewable power sources have even further ramped up interest in projects like the California Valley Solar Ranch. Not surprisingly, conservation groups have responded to this rush toward renewable energy by stepping up their own campaigns to protect treasured natural areas that they see as threatened by renewable energy development. These organizations have found a staunch ally in Senator Dianne Feinstein, author of the 1994 California Desert Protection Act, which gave heightened federal protection to some nine million acres of parks and wilderness in the Golden State.[28]

Watching as the US Bureau of Land Management accepted proposals for renewable energy projects on large parcels of federal land in her state, Feinstein introduced new legislation in 2009 whose goal, in her words, was "to balance the need to protect the California desert while encouraging development of renewable energy projects on suitable lands." This bill, reintroduced in subsequent sessions of Congress, proposed to create two new national monuments, add tens of thousands of acres to existing national parks, and establish five new wilderness study areas covering a quarter-million acres.[29] It's worth noting that within these areas are the sites of several large solar and wind projects in various stages of planning and development. At least one of these facilities—the 392-megawatt Ivanpah solar project in the Mojave Desert—went into operation in January 2014, despite long-standing objections by some conservation groups that this concentrating solar power facility, with its five-square-mile footprint, would disrupt the habitat of desert tortoises nesting in the area, and despite more recent concerns that birds could be maimed or killed by the solar flux created by thousands of mirrors focusing the sun's heat on thermal reactors atop the facility's three tall towers.[30] As of this writing, no legislation subsequent to the 1994 desert protection law has been enacted, though Feinstein has indicated that she may soon reintroduce a bill that would step up federal protection for California desert lands.[31]

The tensions between conserving natural resources and promoting renewable energy have also reached a high pitch in California state politics. Two successive governors have been outspoken advocates for expediting utility-scale renewable energy projects, but they have also highlighted the need to protect priority conservation values. In an executive order issued in November 2008, then-governor Arnold Schwarzenegger acknowledged that the "deployment of new renewable energy technologies across the state will require utilizing new areas of biologically sensitive land." To expedite this controversial allocation of the state's land reserves, he spearheaded the creation of a roadmap that would prioritize areas for renewable energy development on federal, state, and private lands, coordinating between state agencies and the US Department of the Interior. This initiative was ambiguously dubbed the Desert Renewable Energy Conservation Plan, reflecting its dual commitment to promoting renewable energy and protecting California's natural, historic, and cultural resources.[32]

To streamline a regulatory process involving multiple state and federal agencies, Schwarzenegger needed someone who was skilled at negotiating among powerful government stakeholders, adept at engaging the public, and utterly committed to advancing the state's renewable energy goals. He found that man in Michael Picker. A resident of Sacramento for thirty-five years, Picker had worked for Jerry Brown during his first stint as governor in the early 1980s, creating tough controls on toxic substances. Later he led efforts to revive Sacramento's moribund downtown neighborhoods and business district. In Michael's own words, he was seen as an "effectuator."

When I met with Michael in Sacramento's elegant state capitol building in the spring of 2013, he was working for Governor Jerry Brown, continuing the job he had started for Schwarzenegger. Greeting me in jeans and a casual plaid shirt, he led me to a small office in the governor's suite where glossy photos of utility-scale solar plants hung above shelves densely packed with dog-eared reports. On another wall was a "No to Nuclear" poster from an early Jerry Brown campaign. As he laid out his long-term goal and the steps he was taking to get there, it became clear that this was not a man to be tethered by caution or convention.

Michael believes that by 2050, California should be able to rely on renewable energy for 80 percent of its overall electricity needs—about four times the contribution that sources like solar, wind, and geothermal energy supply to the state today. Efficiency measures will help achieve that goal, holding electricity demand steady during an expected period of modest growth in California's population. In pursuing this vision, Michael foresees a level of consumer control over the generation, storage, and use of electricity unprecedented in the history of America's utility-dominated power sector. "Most utilities look at their land as their feudal territory and their customers as serfs," he says, dipping into the radical prose of his early activist years. "They have a hard time conceiving of themselves as enabling people to actually take more control." In Michael's forecast of California's electric future, people will generate at least some of their own power through rooftop and community-based solar installations. They will time the use of their dishwashers, pool pumps, air conditioners, and washing machines to take advantage of off-peak electric rates. And their electric vehicles will provide power storage for the grid along with clean mobility: they will charge up when grid-supplied power is abundant, and their batteries will feed electrons back to the grid when there's a need to balance the intermittent power coming from solar plants and wind farms.

Rooftop solar generation and other distributed energy systems are only part of Michael's vision, however. To achieve 80 percent reliance on renewable energy by 2050, large-scale generating stations are essential, in his view, and the Desert Renewable Energy Conservation Plan is a key means of identifying appropriate sites for these projects. Under the plan's preferred alternative, roughly two million acres in the Mojave and Sonoran deserts are designated as Development Focus Areas where solar, wind, and geothermal energy projects, as well as related transmission, can be permitted on an accelerated basis. New energy projects will probably end up occupying only a fraction of that land, but the plan's underlying assumption is clear: for California to succeed in making a wholesale shift to renewable energy, major areas—including some biologically sensitive ones—will have to be used.[33]

By the time the draft plan was presented for public comment in September 2014, Michael had already changed jobs. In January 2014 he became a member of the powerful California Public Utilities Commission, appointed by Governor Brown. In December of that year, the governor took another decisive step signaling his support for Picker's startlingly bold agenda: he appointed him to serve as president of the commission.[34]

To Janine Blaeloch and others who want solar energy development confined to built-up areas, the Desert Renewable Energy Conservation Plan is an affront. They see huge, unrealized solar potential in our urban areas and wonder why we can't just focus our efforts there. To Michael Picker and other proponents of the plan, building utility-scale power plants on open land is a much quicker, and often cheaper, way to amass the huge increments of clean energy needed to transform California's energy economy. In a power market increasingly dominated by cheap natural gas, the gap between building a relatively small number of very large, ground-mounted solar plants and aggregating millions of residential PV arrays at a much higher per-unit cost makes a compelling argument for utility-scale solar investment. (See fig. 3 in the introduction.) The clash between these widely divergent worldviews will doubtless occupy renewable energy proponents and conservation advocates in California and many other states for decades to come.

CHAPTER SIX

Tribal Sun

ACCORDING TO THE US Department of Energy, the sun shines much brighter on Native American lands than on the rest of the nation. Tribal territories cover 2 percent of our land area, but about 5 percent of America's "total technical potential" for solar-powered electricity resides in those territories.[1] Looking at the electrons that would flow from fully tapping this resource, the government's data show that tribal sun could provide almost five times our total nationwide power needs.[2] Though this number does not take into account competing tribal preferences for the same land such as other economic enterprises, nature conservation, or the preservation of cultural and historic landscapes, it reveals the sun's extraordinary magnitude as an energy resource.

Native American lands have long been exploited, not for the sun that beams down on them, but for the coal seams and uranium deposits that lie beneath their soils. Some of the biggest coal-fired power plants in the country have been built on tribal lands. Among them are the Four Corners Power Plant near Fruitland, New Mexico; the Navajo Generating Station near Paige, Arizona; and the San Juan Generating Station near Farmington, New Mexico. Native Americans living near these plants suffer from respiratory diseases caused or aggravated by the air pollution that they spew into the atmosphere, yet these same plants—as well as the coal mines feeding them—have created thousands of jobs for Navajo, Hopi, and other tribal members. In Indian communities where unemployment can run as high as 40 percent, these jobs are viewed as economic lifelines despite the risks they pose to the health of workers and nearby residents.

Beyond the individual jobs created by coal mining and power plant operations, coal-related leases and royalties have generated a flow of revenues crucial to keeping cash-strapped tribal treasuries afloat. In 2009 the Hopi Tribal Council felt so threatened by activists' efforts to shut down the 2,250-megawatt Navajo Generating Station that it prohibited environmental groups from operating within its tribal territory. In condemning the activities of both indigenous and non-indigenous environmental groups, the council declared that environmental groups "have manufactured and spread misinformation concerning the water and energy resources of the Hopi Tribe in an effort to instill unfounded fears into the hearts and minds of the Hopi public."[3] More recently, the Navajo Nation Council approved a twenty-five-year lease extension for the Navajo Generating Station.[4] In a separate transaction it spent $85 million purchasing the 33,000-acre BHP Navajo Mine, a massive strip-mining operation that fuels the Four Corners coal plant.[5]

While it has been hard to build political momentum in support of alternatives to coal among the Navajo and Hopi, renewable energy has begun to gain traction in some other tribal areas. One of these is the Moapa Band of Southern Paiutes, a small Indian community in Nevada that occupies 72,000 acres of creosote scrub, yucca, and desiccated grasses. Heading northeast of Las Vegas on I-15, I spotted the two landmarks that make it easy to locate this reservation. First was the Moapa Paiute Travel Plaza, a low-slung stucco building dwarfed by large signs announcing its casino—a dark, low-ceilinged room with a few dozen slot machines—and a smoke shop that features the largest collection of fireworks for sale I have ever seen. Driving a bit farther, I came within view of the second landmark: the Reid Gardner Generating Station, a coal-fired power plant with four belching stacks a short distance off the highway. As I soon learned, this half-century-old coal plant sits on land just outside the reservation, but its impacts on the tribe's primary residential community, just a few hundred yards away, have been devastating.

"I think it's fair to say that most of the Washington, DC, politicians attacking clean-air safeguards don't have the same view out their front windows as the families in my small community of three hundred people," observed William Anderson, then-chair of the Moapa

Paiutes. He was referring to the pollution from Reid Gardner that too often settles on Moapa homes and communal buildings, causing elevated asthma rates, chronic eye infections, and a host of other ailments. It's not just the soot, sulfur, and nitrogen pollution spewing from the power plant's chimneys; it's also the toxic coal ash that blows from the plant's open waste pits on windy days. "There are too many mornings when our elders can't have a morning walk outside, too many afternoons when our kids stay indoors because of bad air."[6]

Anderson told me that he and other Moapa Paiute leaders had been struggling for years to get NV Energy, owner and operator of Reid Gardner, to clean up the plant or shut it down. "We worked with the Sierra Club, Earth Justice, and Greenpeace to voice our concerns about the health issues we're facing." He recalled the 50-mile Cultural Healing Walk that tribal members and supporters staged on Earth Day in 2012, starting at the power plant and culminating at the Lloyd George US Courthouse in Las Vegas. The following spring, Anderson and his allies successfully pressed the Nevada legislature to pass a new law mandating the plant's phased retirement starting in 2014, with the last of its four units to be taken off-line by 2017.[7] And in August 2013 the Sierra Club and the tribe filed a federal lawsuit to require remediation of the soils and groundwater that had been contaminated by the plant for so many years.[8]

Along with combating the hazards of an aging coal plant, the Moapa Paiutes have been equally determined to harvest the solar energy that saturates their ancestral land. By the time William Anderson became chairman in December 2010, the tribe had already struck deals for two very large, utility-scale solar PV installations. Once in office Anderson immediately began working to move these projects forward. In March 2012 he met personally with President Obama. "We're just a minority group out here trying to create jobs and renewable energy through solar," he remembers telling the president.[9] When the administration announced seven fast-track renewable energy projects in August of that year, one of Anderson's solar farms was on the list, and the following June, Interior Secretary Ken Salazar approved a second Moapa solar installation.[10] The developer of one of these

projects then nailed down an agreement with the Los Angeles Department of Water and Power (LADWP) as the guaranteed buyer of
250 megawatts of solar-generated electricity over twenty-five years—a
deal worth about $1.6 billion that the utility expects to supply enough
power for roughly 93,000 Los Angeles households.[11] This very large
power purchase will bolster the Moapa Paiute economy while bringing the LADWP much closer to fulfilling Los Angeles's declared goal
of being 100 percent coal-free by 2020.[12]

Land-rich but cash-poor, the Moapa Paiutes will be a landlord
rather than a substantial owner of the solar plants built on their territory. The terms of the contracts remain confidential, but there are
allusions to profit-sharing arrangements supplementing the long-term
lease payments. In any case, all seem confident that solar power will
bring the community a sustained income stream unlike any it has
experienced in the past. Just after the agreement was signed with
LADWP, one close adviser could barely contain her excitement: "This
is life-altering for the tribe! Its economics will be forever changed,"
she told me. As an added benefit, the Moapa Travel Plaza will be
connected to the grid for the first time, freeing it from the polluting
and unreliable diesel generators that have been its only power source.
And it seems fitting that Moapa solar power will be delivered to Los
Angeles via the same transmission line that has supplied the city with
coal-fired electricity from the Navajo Generating Station, across the
border in Arizona. The LADWP has already announced its commitment to shed its 21 percent ownership share of this coal plant by 2015.[13]

—〰—

Other tribal communities across the Southwest are embarking on
modestly scaled solar projects. In Sacaton, the impoverished capital
of the Gila River Indian Community just south of Phoenix, a few
neighborhood streets and a schoolyard are now lit with solar-powered
streetlights funded by the federal government. Oakland-based GRID
Alternatives reports that it has brought solar power to more than one
hundred families in twenty tribal communities since 2010.[14] And in

a remote corner of the Navajo Nation where most people still live without electricity or running water, the Shonto Economic Development Corporation has begun installing residential PV arrays and has brought solar power to an off-grid school.[15]

While solar installations have begun to appear on schools, community buildings, and residential rooftops in many tribal areas, utility-scale solar power plants are progressing more slowly and are much harder to trace. When I met with utility officials of the Gila River Indian Community, I was told that potential sites for a midsized solar farm have been identified and non-disclosure agreements have been signed with a few prospective developers. At the same time, I was given a list of reasons why a utility-scale solar project may not happen. Perhaps most fundamentally, it's not clear that solar power—today, at least—will find a buyer given the availability and abundance of cheap power from coal, gas, nuclear, and hydro plants in Arizona. The utility's chairman John Lewis added that some tribal members might balk at tying up thousands of acres of tribal land with one or more projects whose purpose is to export energy off-reservation, much as coal and uranium have been extracted over so many decades for the benefit of outside consumers. That same land, he said, might be used to build new housing for returning tribal members who can help the community strengthen its economy while building a stronger middle class. He also pointed to higher-value investments that are less land-intensive, such as the lavish Wild Horse Pass Hotel and Casino that his tribe has developed on a few dozen acres along Interstate 10, the main highway leading south of Phoenix. The profits are substantial, the job opportunities are plentiful, and the acreage has been very modest compared to the land needed for a utility-scale solar farm.[16]

In trying to track down other plans for utility-scale solar installations, I found it very difficult to get general confirmation, let alone specific details, about pending projects. In some cases tribal officials alluded to the sensitivity of pending negotiations, but I also sensed a suspicion toward outsiders that was well described by Phil Bautista, director of American Indian solar development for SolarCity, America's leading rooftop solar installer. Part Aztec and part Chippewa, Phil was hired by SolarCity to develop small-scale installations serving communal needs—tribal offices, daycare and senior centers, housing projects, water pumping for farm irrigation—rather than utility-scale solar parks that would be selling their power to off-reservation end-users. Even so, he knew he would bear a heavy burden of persuasion as the sales representative of a non-tribal company based in the Silicon Valley. "It's like going into a whole different country; it's like going to Haiti or Africa," he tells me. "We are a nation within a nation, so I've got to get them to trust us."

When Phil approached SolarCity in the spring of 2013, he brought with him four decades of activism on Native American issues. Along with negotiating water, hunting, and fishing rights for different tribes, he founded a Native American school in Milwaukee, Wisconsin, after becoming frustrated with the disregard of Indian culture and history that his children were encountering in local public schools. To fund the school, he and his team won approval to build the Potawatomi Hotel and Casino, one of the first Indian-owned operations of its kind in the country. The enterprise and the school have been wildly successful, he says. More recently he ran a business selling air purification equipment to tribal casinos. In hiring Phil, SolarCity's management team knew his business acumen as well as his negotiating skills would be boons to the company.

Bringing solar energy to tribal communities is a cause Phil believes in from an environmental perspective, but he also sees an economic opportunity that shouldn't be squandered. "We, as Indian people, should be in the forefront of renewable energy and trying to help our Mother Earth," he tells me. "Way back when, they were rounding us up and putting us on wagons and trains and taking us to desolate areas in the desert, hoping we were all going to die off, thinking it's all worthless land." Now he celebrates in the prodigious renewable energy potential of this territory. "Let's use it!" he says. He approaches tribal council members knowing that the most important thing he can do is inspire their trust. He is keenly aware of the bitterness that many tribal leaders feel about the undervalued leases that signed away tribal mineral rights to outside coal giants like Peabody Coal, and that allowed uranium companies to riddle the Native American landscape with more than a thousand open-pit and underground mines. More recently, he has heard about solar developers selling inferior products to tribal communities and then failing to maintain them. "You have to look at our history, five hundred years of getting ripped off and lied to and sold bad things."

Phil wasn't at liberty to discuss the specifics of the solar installations he has been working on. Little more than half a year since his work began at SolarCity, he was still in the early stages of shaping these projects. In general, though, he told me that the deals he is making are based on power purchase agreements with no upfront payments and predictable electric rates that are lower than those offered by the utilities now serving tribal communities. SolarCity has helped pioneer this economic model with residential and commercial customers across fifteen states.[17]

Ken Duncan Jr., energy coordinator for the San Carlos Apache Tribe, was able to be more forthcoming about the solar projects he has been working on. I met Ken and Phil on the same chilly December day, at an energy summit hosted by Ken's tribe at the Apache Gold Casino and Resort in Globes, Arizona, about 90 miles east of Phoenix. Initially funded by the US Department of Energy's Office of Indian Energy, these annual gatherings are part of a broader DOE effort to expand tribal access to electricity, introduce more sustainable sources of power, and build tribal members' skills in developing and implementing new energy strategies.[18] Ken was hired after the San Carlos Apache planning department had completed a DOE-funded strategic energy plan that identified the need for a point person to focus on renewable energy and energy efficiency projects. Trained in political science at Arizona State University with an additional degree in computer electronics, he had been developing the commercial real estate holdings of the Salt River Pima Maricopa Tribe, just outside Phoenix, when he was offered this job.

Phil and Ken are a study in contrasts. Phil came to the summit wearing a large silver medallion at the neckline of his bold turquoise shirt, with his gray-white hair pulled back into a long, neat braid.

Ken, youthful with his buzz-cut black hair, opted for a muted, button-down shirt and striped tie. Where Phil's words carry the passion and indignation of a veteran tribal rights champion, Ken is a cool, measured pragmatist. He speaks calmly about the technical issues and economic trade-offs involved in the projects he is advancing. His language is peppered with terms like "financial self-sufficiency," "cost-stabilization," and "production-based incentives."

One of the San Carlos Apache Tribe's first solar investments, a 3.5-kilowatt PV array, powers the broadcast signal of KYAY, the tribe's radio station. KYAY, I learned, translates roughly into English as "Wow!" That's not a bad way to describe the effort it took to get this station up and running. Located on a hilltop with no access to the grid, the KYAY radio tower relies on batteries to store enough solar-generated electricity for round-the-clock broadcasting. It was built with funding from a nonprofit environmental group called the Grand Canyon Trust. The KYAY studio is powered by a second, 7.1-kilowatt array, about the size of a large residential solar installation. Mounted on a specially built carport outside the studio, its $40,000 cost was covered by a grant made under the Obama administration's economic stimulus bill, the American Recovery and Reinvestment Act.

Along with airing traditional music, sports, and local news, KYAY provides valuable health and safety information during power outages like the one that hit many of the tribe's twelve thousand residents during my brief visit.[19] The blackout had begun the previous evening, when a power line came down in a remote section of the antiquated grid that serves the reservation. No more than thirty attendees were expected at the tribal energy summit, so I was surprised to find triple that number streaming into the room: men and women of all ages bundled in overcoats and heavy sweaters. Relatively few had come to discuss the tribe's plans for energy development, though. They had learned from KYAY that a free hot lunch featuring traditional acorn soup would be served. The meal, along with a live performance of music and dance led by Ken's father, a revered San Carlos Apache entertainer and storyteller, made the casino's bare-bones conference room a welcome refuge from the tribal members' dark, unheated homes.

Much more ambitious than KYAY's two small solar arrays is the Apache Gold Casino and Resort's planned solar energy project, a series of ground-mounted PV installations that will provide about a fifth of the resort's power needs. What excites Ken about this project isn't just its scale, but its ownership. Rather than signing a power purchase agreement with a third-party owner like SolarCity, the San Carlos Apache Council voted to take possession of the $3.6 million project, leveraging funds made available through the settlement of a long-litigated tribal water rights dispute with various federal, state, and local agencies. The tribe was able to lock in production-based solar incentives offered by Arizona Public Service (APS), the investor-owned utility that serves the western edge of the reservation, including the casino. Under this arrangement, APS will pay the tribe a per-kilowatt-hour subsidy that is expected to approach $2.4 million over twenty years. This subsidy, combined with the substantially reduced size of the casino's electric bills, should guarantee the tribe's five planned solar arrays a positive cash flow throughout the life of the project.

Beyond securing affordable power, Ken has made sure that the casino solar project will yield at least a small number of ongoing jobs for tribal members. Typically, solar developers like the company that the tribe has hired to build the Apache Gold solar arrays remain responsible for operating and maintaining their equipment over a multiyear period. Ken felt this would be a lost training and employment opportunity, so he negotiated a very short transition to local control of the project. Just a year after the project goes live, San Carlos Apache trainees will assume primary responsibility for daily operations. Maintaining the KYAY installations has provided good basic training to tribal technicians; work on the casino project will consolidate those skills.[20]

—⁂—

The Duncan family dance troupe that performed at the San Carlos Apache tribal energy summit is called the Yellow Bird Indian Dancers. Ken's brother Tony, in addition to performing with that group, has his own traditional music ensemble, Estun-Bah, whose recent album is titled "From Where the Sun Rises." Through the mesmerizing flute melodies he plays, Tony conveys the tight connection to nature-rooted folkways that still echo through modern tribal culture. "For Apache people, we believe that the Eastern doorway is where all life begins," he explains of his album's thematic motif. "In the morning, when you wake, you face the East, and you say your first prayers."[21]

Others refer to the obligation to sustain the earth's integrity, not just for those living in the present but also for those yet to be born. Winona LaDuke, twice Ralph Nader's vice presidential running mate, speaks of our obligation to consider the impacts that our actions and inactions today will have on the Seventh Generation, as the future is sometimes envisioned in Native American lore. "We need to create a way of life where community is not forced to cannibalize their mother in order to live," she urges.[22]

Solar power may offer a more assured way to honor future generations than the coal mines and coal plants that have torn up tribal lands, polluted the lungs of so many thousands of tribal members, and altered the broader course of nature around the globe. At the same time, leaders like William Anderson want to make sure that their tribes' embrace of this relatively clean technology doesn't jeopardize the natural legacy that their tribes are bound to honor. Protecting the primeval desert tortoise has been one of his top priorities as he has moved utility-scale solar power forward on Moapa Paiute land.

Indigenous to the Mojave and Sonoran deserts of Southern California, Nevada, and Utah, the desert tortoise's numbers are estimated to have decreased by 90 percent since the 1950s, reaching a current population of about 100,000 creatures.[23] On the 2,000-acre site of the Moapa solar project serving LADWP, a survey in 2010 estimated that 23 to 103 desert tortoises lived in areas that would be disturbed during construction and operation of the facility. Cooperating with the Bureau of Indian Affairs and the US Fish and Wildlife Service, the Moapa development team agreed to relocate all of the tortoises whose habitats were threatened by the project. Before construction began, the tribe set aside 6,000 acres of suitable tortoise habitat as a permanent conservation area, and the project developer, First Solar, earmarked funds for five years of biological monitoring. A total of seventy-five tortoises were rounded up at the project site, and all were fitted with transmitters. According to the trained tribal members who are tracking their movements, none of the tortoises died during the relocation and only one has died since—from a coyote attack.[24]

Wildlife biologist Ileene Anderson at the Center for Biological Diversity led me through some of the risks involved in resettling desert tortoises. First, the creatures have strong homing instincts that often send them on long journeys back to their native turf, during which they are vulnerable to predators as well as the relentless desert sun. It therefore makes sense to fence tortoises *in* to translocation zones, she says, rather than fencing them *out* of their native areas. (So far, the Moapa have not fenced in their tortoise conservation habitat, which is 10 miles from the solar project site.) She also recommends making it as easy as possible for tortoises to re-create their home social contexts. "They have these social hierarchies and they know who their neighbors are," she explains. Moving tortoises in clusters rather than in isolation improves the odds of their adjusting to new surroundings. "When a tortoise is out roaming around, freaking out about 'Oh my God, I just moved to this new place,' it helps when the tortoise that it runs into next is an old neighbor."[25]

Protection of tribal sacred places and cultural landmarks is another concern that has occupied the neighbors and developers of utility-scale solar projects, and no one is more passionate or vehement on this subject than Alfredo Acosta Figueroa, founder of La Cuna de Aztlán Sacred Sites Protection Circle. Alfredo's father was a Yaqui Indian from Sonora; his mother was from the Colorado River Indian Tribes reservation. Now past eighty, Figueroa is an activist down to his bones, beginning with his organizing efforts for the California farmworkers' movement in the 1960s, running through his opposition to a proposed nuclear power plant outside San Diego in the 1970s, and followed by a successful campaign to block a nuclear waste dump in California's Mojave Desert in the 1990s.

Figueroa is also a self-taught student of Aztec/Mexica history, which he believes to have deep roots in the Palo Verde/Parker Valleys of the Lower Colorado River Basin. Based on his examination of tribal codices, his intimate familiarity with geoglyphs and petroglyphs (landscape-embedded designs and rock engravings), and his study of the area's mountains and other landscape features, he is utterly certain that this is the site of Aztlán, the ancient homeland of the Aztec/Mexica people and the touchstone of its creation story. "Today, like the ancient Phoenix Bird, Aztlán is rising from the ashes of obscurity," he proclaims.[26] While many regard Atzlán as closer to myth than history, there is a wealth of artifacts in the area that Figueroa and his allies are dedicated to preserving. Unfortunately, some of these collide geographically with plans for several of Southern California's biggest solar power projects.

Figueroa's group, La Cuna, has filed a number of lawsuits against the US Department of Interior, other government agencies, and solar project developers, claiming that several utility-scale solar power stations were approved without due regard to the damage they would cause to Native American cultural and religious sites. In the media, the sparring surrounding La Cuna's claims has been intense, with Figueroa's allies maintaining that some threatened geoglyphs go back thousands of years and developers countering that the contested formations, in some cases, are no more than a few decades old.[27] La Cuna also asserts that there has been little consultation with tribal groups, in violation of the National Historic Preservation Act and other federal laws. Government officials insist that adequate consultations with multiple tribes have taken place, a position that the courts have upheld in at least two cases.[28]

The California projects contested by La Cuna are on non-tribal public and private property, but the controversies surrounding them mirror the difficulties that solar developers often face as they look for sites large enough to accommodate utility-scale power plants on tribal lands. Solar fields may tread lightly on our global climate—a subject we will explore in chapter 8—but there's no denying the more localized impacts they have on the lands they occupy. Setting aside hundreds, and sometimes thousands, of acres for solar energy can weigh heavily as tribal communities reckon with the possible impacts of these projects on historical landscapes and cultural artifacts. Figueroa states it stridently when we discuss the battles he is currently waging. "Any time you destroy certain sites, you're just a continuation of the Spanish Inquisition," he says. "We are the guardians of Mother Earth."[29]

Rebecca Tsosie, a law professor at Arizona State University, is keenly attuned to the tensions between present economic interests and more enduring environmental and cultural values. Half Yaqui by descent, she is a widely published authority on tribal energy law who deeply believes that the time for "indigenous cultural sustainability" has come. Reflecting on the tensions between coal and renewable energy development on tribal lands in the West, she says: "There is an infrastructure set up for coal-fired power plants, the energy grid serves it, and the long-term contracts have already been articulated with the coal companies. So the economics, the history—it's all supporting continued development of a resource even though we know, in an era of climate change, that it's not a sustainable energy resource for the future."

Tsosie proclaims herself to be "a hopeless idealist." Amidst all the conflicts and contradictions, she senses the emergence of a renewed commitment to "traditional ethics of the land." She sees these strengthened values creating new bonds between communities as diverse as the Canadian Inuit, whose native lands may be among the first coastal areas to succumb to sea-level rise, and inland nations like the Navajo, who are equally important players in the unfolding drama of climate change. A well-traveled speaker on indigenous people's rights, she often addresses non-indigenous audiences as she seeks to build a shared ethos of sustainability. "Non-Indians come up to me after a lecture saying, 'You know what? My family has been here for X number of generations. This is my land too. And I totally want this land to survive, to thrive. . . . Let's have this sustainable energy development and let's talk about what it means to be part of the community and part of the land.'"[30]

Given the dire economic conditions facing many Native American tribes, it will take more than a new embrace of traditional values to bring about a wholesale switch away from polluting fossil fuels, especially in areas where those fuels are the primary generators of jobs and tribal revenues. Yet, as the economics of technologies like solar and wind continue to improve, we will likely see greater numbers of pragmatists like William Anderson and Ken Duncan Jr. joining "hopeless idealists" like Rebecca Tsosie in their embrace of more sustainable energy choices.

CHAPTER SEVEN

Focusing on Tonopah

MOST AMERICANS ASSOCIATE SOLAR energy with the photovoltaic panels on their neighbor's roof or the PV arrays they have seen on a local public school or sports arena. Photovoltaic technology does, in fact, supply the majority of our solar power today: 18.3 gigawatts of it by the end of 2014,[1] equivalent in installed capacity to several dozen US coal plants.[2] Almost half of that solar power comes from PV panels on our homes, businesses, and public buildings. Utility-scale PV installations provide the rest.[3]

While PV is the solar technology of choice, some utility-scale solar power producers have begun using concentrating solar power, or CSP, to harness the sun's energy. Rather than converting sunlight into power, CSP relies on the sun's heat to generate electricity. It is therefore sometimes referred to as solar thermal power. Several CSP technologies are now in use, all of them relying on mirrors to concentrate the sun's heat. Some CSP plants have large fields of mirrored parabolic troughs. More captivating, though, are the solar power towers that focus thousands of giant mirrors, or heliostats, on a receiver mounted hundreds of feet in the air. The parabolic trough and power tower plants use sun-generated heat to produce the steam that turns electricity-generating turbines similar to those used in coal and gas-fired plants. Much less commonly used, sun-tracking disks resembling satellite dishes on steroids heat a gas such as hydrogen, whose expansion creates the pressure to drive a closed-cycle engine.[4]

I visited my first solar power tower on a high desert plateau in Nevada, halfway between Las Vegas and Reno. The Crescent Dunes Solar Energy Project lies a few miles outside Tonopah, a small town made legendary by the silver miners who flocked there at the turn of the twentieth century.

There are two ways to get to Tonopah from Las Vegas. The lowland route runs through the Amargosa Valley, past the 1,350-square-mile Nevada Test Site, still off-bounds to the public because of ongoing military activities and the radioactive remains of nuclear detonations dating back to the 1950s. I chose the highland route even though it was a bit longer. On the Great Basin Highway, I drove north through a hundred miles of arid upland terrain whose desolation was broken only by occasional small herds of cattle. From my previous research into wind power, I knew that a transmission corridor may soon run through this wilderness expanse, carrying high-voltage electricity from giant wind farms in Wyoming, now in various stages of permitting, down to greater Las Vegas and Southern California. I stared out the car windows at all this open land, much of it relentlessly flat, and couldn't help imagining the possibilities for building a whole new network of solar fields that could piggyback onto the planned power lines.

Hours later, after traveling another 150 miles on the fittingly named Extraterrestrial Highway, I arrived in Tonopah, population 2,300. Just beyond a few weather-beaten motels and boarded-up storefronts, I found the small stationery store where Bobby Jean Roberts, advertising manager of the *Tonopah Times-Bonanza*, has her office. Like the dusty bound newspaper volumes that surround her, Bobby Jean is a repository of Tonopah's history of hopes and disillusionment. First it was silver mining. All that's left of this once-bustling industry is a barren hillside riddled with old mine shafts, hoist houses, and rusting rail cars. Then, during World War II, fighter pilots training at a nearby airbase brought a brief uptick in commerce to the town. Later, weapons testing at the Nevada Test Site triggered a mini-motel boom until workers were transferred to separate housing.[5]

Given Tonopah's uneven history of industries coming and going, it's not surprising that local residents were leery of SolarReserve's proposal to build a solar power tower on the high-desert plateau 12 miles outside town. Bill Roberts reflected this distrust in a *Times-Bonanza* column. "Tonopah residents . . . have been burned so many times it is not uncommon for us to wear fire-retardant underwear when we hear that the next great plan to save Tonopah will be presented at a town board meeting."[6] Yet in spite of all their past disappointments, Bill and Bobby Jean were prepared to give the Santa Monica–based energy company the benefit of the doubt.

I visited Crescent Dunes when the $1 billion project was nearing the halfway point in its thirty-month construction cycle. Its round concrete tower, standing 640 feet tall, was visible from miles away as we drove down to the desert floor from Tonopah's perch astride a gathering of gnarled, barren hills. My host was Brian Painter, Solar-Reserve's construction site manager and a veteran of energy projects ranging from gas turbines in Thailand to oil-fired power generation in South Korea. Crescent Dunes, the first concentrating solar power tower with on-site heat storage in America, was a pioneering adventure that he was excited to be part of.

On our way out of town, Brian and I pass a concrete batch plant, just a block or two off Tonopah's main street. Harris Fence and Concrete has enjoyed a big jump in sales as a supplier of concrete to Crescent Dunes, but other local businesses have benefited as well. Local motels have seen their reservation books fill up with construction workers and specialized tradespeople; local stores have enjoyed a spike in receipts; and the recently revived Mizpah Hotel, a century-old brick beauty, has hosted visitors from around the world who have come to learn about SolarReserve's leading-edge technology for capturing and storing solar energy.

In his soft-spoken Canadian way, Brian radiates enthusiasm about his project. "I think this is the real deal," he tells me as he steers his pickup toward the still-distant tower. "Tonopah is the center of the earth as far as the solar industry goes." SolarReserve CEO Kevin Smith has regaled him with stories of conversations he has had with energy leaders in places as far-flung as Germany and South Africa. "Hey, how's Tonopah going?" they ask. From his boss's global travels to his own immersion in the technology, Brian is a confirmed believer: "When we get this one running, it will become the leading edge of how to make energy."

Aerial view of Crescent Dunes Solar Energy Project, a concentrating solar power facility near Tonopah, Nevada. (Photo credit: SolarReserve, LLC)

As we pass through the security gate, I begin to fathom just how daring a proposition it is to build a CSP plant large enough to supply the electricity needs of seventy-five thousand Nevada households when operating at its peak. The poured concrete foundations for many of the solar reflectors, or heliostats, are already in place, radiating outward in evenly spaced, concentric circles from the receiving tower. Soon more than ten thousand heliostats, each one 37 feet wide and 34 feet tall, will be pole-mounted on these triangular pads. The outermost circle is 1.75 miles in diameter, and the whole constellation covers 1,600 acres of pancake-flat, sandy desert soil. Brian tells me it will take 84 miles of access roads to service all of them, and a permanent cleaning crew to keep their mirrored surfaces relatively free of windblown grit.[7]

To capture the sun's heat, each heliostat must be precisely aimed at a receiver that sits like a lighthouse beacon atop the plant's tall tower. Every ten seconds, a GPS-guided double-axis tracker adjusts each heliostat's position, targeting the sun's rays at molten salt running through pipes in the receiver's outer wall. I learn that the software used to control the trackers was developed by Aerojet Rocketdyne, a company that has its roots in the Cold War arms race and even today draws much of its business from the manufacture of rocket propulsion and guidance systems for the military. At least this small fraction of the company's output is yielding ploughshares rather than swords.

Heated to more than 1,000°F, the molten salt flows in large steel collector pipes down to the base of the concrete tower. From there it can be channeled to a heat exchanger that uses the captured energy to create steam for a power-generating turbine—*if* there's an immediate demand for electricity. Alternatively, the super-heated fluid can be pumped into a 3.6 million gallon stainless-steel storage tank just a few dozen feet from the tower.

The use of molten salt, a combination of sodium nitrate and potassium nitrate, is the innovation that inspires CEO Kevin Smith's greatest excitement about the project. Once heated, the fluid compound

Installation of heliostat mirror at Crescent Dunes Solar Energy Project near Tonopah, Nevada. (Photo credit: SolarReserve, LLC)

retains its heat with minimal losses over many hours, or even days. (Overnight it will shed no more than 1 percent of its heat value, Smith tells me.) This large reservoir of on-site storage capacity distinguishes Crescent Dunes from all but a few other CSP facilities and from all large-scale solar generation based on photovoltaics. While it's more expensive to build and operate a CSP plant with storage, Smith is banking on his power plant's flexibility to store heat and then use it to generate electricity when it will command the highest price on the wholesale market. During the cooler winter months, for example, electricity demand may spike in the early morning and early evening, when his plant is not drawing energy from the sun. The plant's stored heat can also be converted to electricity on cloudy or rainy days, creating a steady flow of electricity regardless of the sun's intermittency.

Kevin Smith knows that low natural gas prices and the plummeting cost of photovoltaics make CSP a tough sell, but he takes heart in the fact that NV Energy, Nevada's biggest public utility, has purchased all of Crescent Dunes' power for its first twenty-five years of operation. He also acknowledges the pivotal role that the US Department of Energy has played in helping him commercialize SolarReserve's unique brand of concentrating solar power. Without a $737 million loan guarantee from the federal government, Crescent Dunes would not have been feasible given its high capital costs, its unorthodox technology, and the difficulty attracting investors in the years following the 2008 financial market collapse. Those guaranteed funds, together with the 30 percent federal investment tax credit and $260 million in private equity, have allowed the project to move forward. The SolarReserve CEO rebuffs the critics inside and outside Congress who have savaged the federal loan guarantee program as an under-scrutinized handout to renewable energy companies like Solyndra, the failed solar panel manufacturer that, in fact, is one of the few loan guarantee recipients to renege on its repayment obligation.[8] Smith stresses that, in addition to repaying the principal, SolarReserve will provide roughly $300 million in interest over the twenty-two-year course of his company's guaranteed loan.

In a large steel-frame shed near the Crescent Dunes security gate,

I witness the preparations for assembling the heliostats—too large to be transported from an off-site factory. Each heliostat consists of thirty-five mirrors mounted on a steel frame, using glass produced by Rioglass, a Spanish company, at its plant in Arizona. A dozen or so hard-hatted engineers are busy calibrating the assembly-line equipment, readying it for a robust output of 84 units per day on a double production line running two eight-hour shifts over several months. I notice that they are speaking Spanish, and I learn that the contractor for building Crescent Dunes is a Spanish company called Cobra. Kevin Smith tells me that he would have preferred to hire an American firm, but Cobra's experience building a similar CSP facility in Spain, though on a smaller scale, gave it a competitive edge. He was also persuaded by the company's commitment to deliver an operating power plant at a fixed price on an agreed-upon completion date, with liquidated damages for any shortfalls. No US bidder would offer comparable terms for this novel technology.[9]

At least half of the several hundred construction workers at Crescent Dunes are from Nevada and upwards of 95 percent are American, but hiring a Spanish contractor has stirred up a surprising amount of animus among Tonopah residents. Nye County Commissioner Joni Eastley, a prime proponent of the project, senses this resentment among her constituents and apparently shares it, to some degree. "When you go to the grocery store or the post office or even the restaurants and you hear the majority of people are speaking Spanish, that's caused a lot of problems for us," she tells me. We are talking over lunch in the quaintly refurbished dining room at the Mizpah Hotel. A few minutes later, a table across the room fills with half a dozen Spanish-speaking engineers from the construction site. "See those Spaniards down there?" Her gaze directs me to the Cobra crewmembers, chatting comfortably as they settle into their seats. "You see what I mean."

Though uncomfortable with the influx of Spaniards, Eastley makes it clear that local residents, in her view, have not been edged out of jobs because of them. Many have left Tonopah over the years as past employment sources dried up, but among those who have stayed, unemployment is low—somewhere between 5 and 7 percent, she says. Jobs may require some travel—for example to the Tonopah Test Range 30 miles southeast of town, or to Round Mountain Gold Mine 60 miles to the north. But most people feel secure in their jobs and aren't likely to give them up for temporary construction work. "If you live here and you want to work, you can have a job. If you're not working, there's a reason. It's because you don't want to work, or because you can't pass a drug test, or because you're ill or disabled."

Once SolarReserve begins recruiting for permanent jobs at Crescent Dunes, Eastley expects the level of local interest to increase. Half of the plant's forty-five staff positions are expected to be filled by Tonopah residents—a big lift for a town that has fewer than five hundred people in its workforce. Some functions requiring specialized skills may call for outside recruits, but there will be ample work maintaining the site. In this high-desert environment with its sandy soils and occasionally strong winds, cleaning crews will stay busy year-round keeping more than ten thousand heliostats free of sun-blocking grit. If a few dozen townspeople end up employed at the site, Eastley will be satisfied.

Halfway through lunch, Eastley and I are joined by Tonopah town manager James Eason. Clean-cut, upbeat, and bristling with boyish charm, he arrives on a break between coaching two high school basketball teams. He and Eastley have waged a long campaign to lure solar developers to the open desert land around Tonopah. From the start they knew that the area bordered air space reserved exclusively for military testing and training, but they were confident that wildlife concerns at local elevations topping 5,000 feet would be minor compared to those being managed at a number of less lofty solar sites. When Kevin Smith decided to locate his company's CSP project near town, they were delighted. "We've always said we want base hits," Eason tells me. "SolarReserve turned out to be a home run." Along with the expected jobs, Eastley and Eason happily embrace the quickened pace of local business during the plant's construction, along with the expected boost to local tax revenues, estimated at $40 million over the first ten years of the plant's operation. They also take heart in the $300,000 Community Endowment Fund that SolarReserve has created for local infrastructure investments and student scholarships.

As enthused as these two politicians are about the benefits of bringing solar power to the area, both make sure I understand that they are far from embracing renewable energy as a substitute for other sources of electricity. "If somebody wants to site a nuclear power plant out here, I'll be the first one to promote it," Eason proclaims. He says the same about coal. "You cannot be reliant on one single energy source," he says. Caustic about those who have blocked the long-term disposal of nuclear waste at Yucca Mountain, in her jurisdiction, Eastley chimes in with a broadside against environmentalists who have raised concerns about the SolarReserve project. "You want to talk about the hypocrisy of environmentalism?" she asks. "The most ridiculous one we heard was, 'We don't want the tower down there because it's going to interfere with the viewshed.'" Her exasperation builds. "WHAT viewshed? There are literally tens of thousands of miles of desert out here! We're going to take a pinprick of one area for a renewable energy project and you're going to complain that it interrupts your *viewshed?*" Eason adds his own pet peeve: the Center for Biological Diversity's call for a study of the project's possible impacts on the pale kangaroo mouse, whose habitat runs through the Crescent Dunes area. They couldn't find anything more serious to worry about, he says.[10]

Seeking a more nuanced perspective on the power plant's environmental impacts, I drop by the US Bureau of Land Management's Tonopah Field Office. Since SolarReserve's project sits on leased federal land, the BLM has taken the lead in evaluating its environmental impacts. In a small, ranch-style building on the upland edge of town, I meet Field Manager Tom Seley; he is in charge of about 6.1 million acres in BLM's Battle Mountain district. In a calm, matter-of-fact manner, he guides me through the issues his office has examined.

Anticipated water use at Crescent Dunes is estimated at 600 acre-feet per year, or about 3 percent of Nye County's groundwater consumption for irrigation, mining, and milling.[11] Some of this water will be supplied to the plant's hybrid air-and-water cooling system; the rest will be used to clean dust and desert sand from the heliostats' glass surfaces. ("Ten thousand windows to be washed," Bobby Jean Roberts commented when we met at the *Times-Bonanza*. "Talk about the job that never ends!") The BLM has made sure that these water needs will be met through the purchase and retirement of irrigation rights in the general area. This measure has been contested by the Center for Biological Diversity, however. The Center claims that retiring water rights 10 miles from the project site will not prevent more localized negative impacts on springs and groundwater.[12]

Earthquake protection is another concern that Tom has attended to. He has made sure that the compaction and geotechnical characteristics of the soil were carefully studied and the tip-over calculations for the power tower were rigorously prepared. Keeping a sixty-storey concrete monolith stable and vertical is no small matter in a state that, he reminds me, ranks just behind California as "the most seismically active in the lower forty-eight."

The pale kangaroo mouse is the first animal Tom mentions when running through the wildlife species that may be affected by the project. As this small rodent's local numbers and distribution are not fully known, he tells me that SolarReserve has committed $200,000 to a study "so we can make better management decisions going forward." What *is* known is that nearly 30,000 acres of potential or suitable habitat for the pale kangaroo mouse exist in the alkali sands and desert scrub surrounding the solar plant.[13] Though related to the giant kangaroo rat, a focus of concern at the California Valley Solar Ranch, it is not listed as threatened or endangered.[14]

Next on Tom's list are golden eagles. One occupied nest was spotted 8 miles from the project area; another inactive one was found 4 miles from the site. "Their greatest food supplies are house cats and small birds," he tells me, "so they'll nest closer to town." In any case, the project's sandy, treeless surroundings are a poor habitat for eagle roosting and nesting.[15]

The greatest risk to birds at Crescent Dunes is "flux," the intense heat surrounding the receiver, with nearly 13 million square feet of mirrors beaming the sun's rays directly at it. While eagles in the area may be few in number, the Center for Biological Diversity has taken the BLM to task for failing to assess the broader hazard to migrating raptors and passerine species flying through what the advocacy group calls "the death ray zone." In its environmental impact statement, the BLM cited a single study conducted in 1986 at a small CSP facility, estimating that 1.7 birds per week were killed by reflected heat at the site. The Center, in its public comments, objected that the agency should have weighed more carefully the possible impacts of flux at Crescent Dunes, given the plant's much larger scale. The group also asserted that a more careful study of bird migration through the area would have made it easier to anticipate how many birds might be harmed or killed at the facility.[16]

While the actual toll on birds at Crescent Dunes is hard to predict, early indicators from another project, the twin-tower Ivanpah Solar Electric Generating System in California's Mojave Desert, have caused mounting concern about power towers generally. During Ivanpah's test phase in the fall of 2013, dozens of birds were found scorched or singed at the site. Estimates of future bird fatalities at the facility range from a low of 1,107 to as many as 28,380 deaths per year.[17]

Chastened by Ivanpah's early experience, the California Energy Commission issued a proposed decision in December 2013, freezing plans for the twin-tower Palen CSP plant in Riverside County. The following September, just before the commission was to release its final decision, BrightSource Energy and Abengoa—Palen's developers—announced that they were abandoning the project. Ongoing concerns about birds were a factor in this decision, though the two companies also feared that they might not complete the facility in advance of the deadline for obtaining the 30 percent federal investment tax credit, set to expire on December 31, 2016. Thereafter they would only qualify for a 10 percent credit—a potentially crippling financial blow to the project.[18]

SolarReserve's Kevin Smith is disturbed by the news accounts that have latched on to still-small numbers of bird fatalities at Ivanpah, distorting the actual severity of the damage caused by flux. "It somehow seems that the drama of a bird getting singed with solar energy is more press-worthy than a bird just flying into a glass building," he says. "It's completely out of perspective." He cautions that power tower flux needs to be viewed in the broader context of the much greater dangers we humans pose to bird populations through the skyscrapers we build, the cars we drive, and the other energy industries we operate. Studies have estimated birth deaths caused by collisions with buildings at 100 million or more per year, and birds killed by motor vehicles at around 80 million annually.[19] "When you compare [power tower flux] to the other risks associated with birds, it's infinitesimally small," he says. "As the guys who do the bird studies say, it's not biologically significant."

Discussing the avian impacts of solar power tower technology with Kevin Smith brings to mind a conversation I had when writing my recent book on wind power. The communications director of a leading wind turbine manufacturer bluntly told me: "Your neighbor's cat is a far more efficient killing machine than my turbines."[20] Reinforcing this point, wildlife author Richard Conniff has written that free-ranging cats kill about 2.4 billion birds each year in the lower forty-eight states.[21] In fairness, though, the favored prey of cats tend to be smaller songbirds, not the larger raptors that have been killed at a number of wind farms and that may be vulnerable to power tower flux.

Vigilance is surely needed to minimize bird deaths at solar power towers, but in weighing this hazard, it's important not to lose sight of the devastation to bird populations and entire wildlife habitats that lies ahead if we fail to rein in global greenhouse gas emissions. We *need* to develop responsible alternatives to the fossil fuels that are placing our global environment in jeopardy. Sited sensitively, power towers may be a valuable part of our renewable energy mix.

—⁂—

Solar power towers have become media magnets, and not just because scorched birds make good news copy. In an era when photo images zip around the world via Facebook, Twitter, Instagram, and other electronic means, there is an unmistakable allure to the perfect geometry of precisely positioned heliostats encircling a brilliantly illuminated desert beacon. The technology is so visually riveting that it inspired Irish artist John Gerrard to mount a 28-by-24-foot outdoor art installation at Lincoln Center in the fall of 2014, using digital imagery to simulate ground-level and satellite views of Crescent Dunes under daytime and nighttime conditions.[22] It's not surprising that the more widely used utility-scale solar thermal technology—parabolic trough CSP—has remained in the shadows.

Parabolic trough CSP was first introduced commercially in the United States in the mid-1980s by Luz Industries, an Israeli company that used this technology when it built nine utility-scale solar fields in California's Mojave Desert. By 1991 the company had filed for bankruptcy, edged out of the market by cheap natural gas prices and the collapse of government incentives for renewable energy investments. Ownership of the Luz solar fields has changed hands since then, but the parabolic troughs continue to produce power today. Rated at a combined total of 354 megawatts, these facilities—known collectively as the Solar Energy Generating Station, or SEGS—have about three times the nameplate capacity of the Crescent Dunes power station in Tonopah. They are a tribute to the durability of a pioneering technology, if not the endurance of the corporation that commercialized it.[23]

One company that has carried parabolic trough CSP into the twenty-first century is Abengoa, a Spanish multinational corporation whose technology investments range from water desalination to electricity generation. In October 2013 Abengoa pushed the start button on the world's largest parabolic trough power station at Gila Bend, about 70 miles west of Phoenix. The 280-megawatt Solana project generates enough electricity for about seventy thousand homes, and it is similar to Crescent Dunes in one important respect: it has adopted molten salt storage as a central feature of the plant's operations, extending its ability to deliver solar power to the grid well beyond the hours when the sun is beaming down on its long rows of mirrored solar concentrators.

Driving out to Gila Bend, I pass through miles of alfalfa. The fields' high borders tell me that this crop is watered by periodically flooding entire fields in defiance of the arid climate and searing heat.

Gila Bend itself is a bleak hamlet with a handful of roadside inns, fast-food eateries, and a welcome sign that proclaims it to be the "Home of 1700 Friendly People and 5 Old Crabs." Today its chamber of commerce website boasts a more upbeat tagline: "Arizona's Solar Capital—The nation's leader in the creation of a modern, renewable energy grid."

A few miles west of town, I turn onto a side road that runs alongside a giant cattle feedlot. The stench of urine is unmistakable and overwhelming. On the opposite side of the road is a tall, slatted fence punctuated only by a guardhouse and entrance gate. There I wait for Emiliano Garcia, general manager of the Solana power plant. A native of Spain, he has spent the past fifteen years with Abengoa, initially working on transmission projects. Since coming to the United States in 2007, his focus has shifted to developing Abengoa's solar assets.

Site preparations for the Solana project were relatively straightforward, as Emiliano describes them. The 1,900 acres purchased for the power plant had been owned by a single farming company that had grown alfalfa, soy, and corn, mainly for cattle feed. Given this heavy prior use, little wildlife remained on the property—just a few dozen burrowing owls. Emiliano brought in an outside firm to capture the owls and build artificial burrows for them outside the construction zone. Now that the plant is complete, some of the owls have been relocated to their original territory inside the perimeter fence.

Built on a rectangular grid, Solana's parabolic troughs are less striking from afar than the thousands of concentrically arranged heliostats at Crescent Dunes. From up close, though, the plant inspires a different kind of awe. As I follow my host's car down a half-mile service corridor to the facility's power block, in the center of the property, the giant mirrored arcs stand at triple the height of my rental sedan, towering overhead on both sides of the narrow access road. Each module, 20 feet long, uses 28 mirrors to create a parabolic aperture of 15 feet, and there are 32,700 of these modules in a jaw-dropping sequence of seemingly endless rows: a glass-and-steel world where nothing else is visible.

When we arrive at the power block, Emiliano takes me into

Parabolic trough reflector at Abengoa Solana Generating
Station, Gila Bend, Arizona. (Photo by author)

the operations office, a small steel-frame building surrounded by
industrial-scale storage tanks, water treatment equipment, power turbine machinery, cooling towers, and lots of mammoth metal pipes.
As he explains the plant's workings, I see how it is like and unlike
Crescent Dunes. Both facilities use conventional steam turbines to
generate electricity, and both use molten salt as a storage medium,
but that's where the similarities end. Instead of running molten salt
through a tower-top receiver, Solana uses a much lighter synthetic
oil to collect the sun's heat. The oil is pumped through two-inch steel
tubes that are suspended in slightly larger glass conduits just above the
mirrored parabolic troughs, raising its temperature to about 740°F.
Then it is carried via a network of insulated collector pipes to the
plant's power block. If there is a ready market for electricity, the plant
operator channels the oil to a heat exchanger that converts water to
steam in a closed cycle that feeds two conventional steam turbines. To
delay power production or spread it over a longer period, the operator
can transfer the oil-borne heat to molten salt, a more stable storage

medium. Later, when power is needed, the stored heat can be used to produce steam.

We drive through the plant in Emiliano's car. He points to six squat, cylindrical tanks; each contains 20,000 metric tons of molten salt. In a yard near the storage tanks, he shows me how the salt arrives—in bulging woven nylon sacks weighing 1,200 kilograms apiece, prominently stamped "ORIGIN CHILE." Using a propane heater, the salt's temperature is gradually raised. "You fill the buffer, melt, fill the buffer, melt, on a continuous basis for six or seven months," he tells me. At 485°F, the crystals melt, and the salt is then kept in a molten state throughout the lifespan of the plant's operation. When used to store solar-generated heat, its temperature can reach 730°F.

As we approach one of the solar fields, Emiliano stops his vehicle so we can get a close look at the trough-shaped reflectors, mounted on a web of crisscrossing steel struts. He points to the hydraulic motors that shift the angle of the mirrors to gather the brightest sun. "We need to track the sun within milliradians," he says. Spread across the three-square-mile complex are five weather stations that collect and transmit the data needed to make these very precise adjustments.

Water is a constraint at Solana, just as it is at Crescent Dunes. Its primary uses are to operate the steam turbines and cool the power plant machinery. Although the power block at Solana has a closed-loop water cycle, about 5 percent of its water is lost to evaporation or needs to be discarded when it becomes too contaminated by natural minerals such as calcium. In all, the Solana plant will consume about 3,000 acre-feet of water per year. This is a tenth of what was previously used for irrigating alfalfa on this parched stretch of farmland, Emiliano tells me.[24] That's the good news. The bad news is that the life-cycle water consumption per kilowatt-hour at a water-cooled CSP plant like Solana is likely to exceed that of a nuclear or coal-fired power plant, and may be more than twice that of a combined-cycle gas plant—though that doesn't account for the very large amounts of water used (and often heavily polluted) when gas is extracted via hydrofracking.[25]

Dry-cooling a CSP plant can reduce water consumption by 77 percent, but it costs more and requires about 8 percent more energy to run a plant of that kind.[26] SolarReserve has struck a balance at the Crescent Dunes power tower, relying on a hybrid cooling system that is expected to reduce that plant's water consumption by as much as 600 acre-feet per year.[27]

While water for cooling isn't needed at PV solar fields like California Valley Solar Ranch, both of these technologies consume water—though in much smaller amounts—for another purpose: cleaning the surfaces of PV panels and CSP mirrors to keep them free of grit.[28] At Solana, Emiliano anticipates that the reflectors will need to be cleaned every other week. He shows me the fleet of spray trucks and brush-bearing, one-armed vehicles, bug-like in appearance, that will perform this job. Given the facility's size, they will seldom stand idle.

—⁂—

Concentrating solar power remains somewhat novel, especially at plants like Crescent Dunes and Solana where molten salt storage allows for flexible dispatching of electricity. Yet even if the technology performs exactly as intended, major questions surround the financial viability of CSP. To build the $2 billion Solana plant, Abengoa landed a federal loan guarantee of $1.45 billion, and it further benefited from the federal government's 30 percent renewable energy investment tax credit. (Granting federal incentives to foreign companies like Abengoa has raised the hackles of some members of Congress, even though America's obligations under international trade law demand non-discrimination on such matters.) Both of these federal programs have an uncertain future given the antispending wave that has swept the US Congress. The loan guarantee program has been singled out for attack by conservative Republicans like House Budget Committee chairman Paul Ryan. When the committee unveiled its 2014 budget proposal, Ryan unfairly pointed to the Crescent Dunes loan guarantee as one of the program's "ill-fated ventures," likening it to Solyndra. Ironically, when Ryan leveled this volley at Crescent Dunes, Solar-Reserve was already well under way with constructing the project and was showing every indication that it would fully pay back its loan in a full and timely manner.[29]

Aside from the uncertainty of future Congressional spending on renewable energy, CSP project developers face intense price competition in the solar energy marketplace. Looking at different solar technologies, the financial advisory and asset management firm Lazard estimates that utility-scale photovoltaic plants produce power at a life-cycle cost of 7.2 to 8.6 cents per kilowatt-hour in areas with relatively high solar radiation. Concentrating solar power installations with thermal storage are assigned a higher cost of 11.8 to 13 cents per kilowatt-hour.[30] According to the Fraunhofer Institute for Solar Energy Systems, though, the life-cycle cost of CSP solar is much higher: from 18.5 cents per kilowatt-hour in areas with very high solar irradiation (such as the deserts of California) up to 28.5 cents per kilowatt-hour in areas with moderately high solar exposure.[31] Even though CSP projects have the appealing feature of on-site thermal storage, it will be hard to attract investors to these projects unless economies of scale and technological innovation make them cheaper to build and operate.

When the California Public Utilities Commission rejected a power purchase agreement for one of two 250-megawatt CSP units at the Rio Mesa power tower project in October 2012, it cited the proposed facility's high price tag compared to that of other solar projects. The second unit at Rio Mesa was also deemed to have "poor value," but the commission approved its power purchase agreement because it sought to encourage the unit's proposed use of molten salt storage.[32] Just a few months later, Rio Mesa's developer announced that it was mothballing the project. Financial strains were among the stated reasons, along with concerns about wildlife impacts, the discovery of hundreds of Ice Age fossils on the site, and Native American claims that the project would destroy their sacred lands. Proposed for a 4,000-acre site overlooking the Colorado River, Rio Mesa was one of the solar plants opposed by Alfredo Figueroa's La Cuna de Atzlán Sacred Sites Protection Circle.[33]

Given the multiple stumbling blocks facing the developers of CSP facilities, it's not surprising that they occupy a relatively small place in America's solar energy constellation. With 2.2 gigawatts in operation by December 31, 2014, CSP accounted for little more than a tenth of our overall installed solar capacity as of that date, and less than a fifth of all utility-scale solar power.[34] Whether the obstacles are financial, environmental, archaeological, or cultural, finding viable parcels of land in the Southwest that are well suited to CSP is proving to be a bigger quandary than solar developers might have anticipated when they started scouting out sites several years ago. Open questions about the toll of power towers on birds have only further complicated the future trajectory of this technology.

On the positive side of the environmental equation, CSP plants significantly outperform solar photovoltaics in generating low-carbon electricity, emitting an estimated 22 grams of CO_2 equivalent per kilowatt-hour as compared to PV's 46 grams per kilowatt-hour—including all the energy burned in manufacturing the CSP hardware, building the plant, and operating it over the course of its expected lifetime. At that level of performance, CSP plants produce less than 5 percent of the carbon emitted by natural gas–fired power plants and 2 percent of coal plants' carbon pollution.[35]

Of more immediate practical consequence to utilities that are in the market for renewable energy, the option of having thermal power storage at a CSP plant is a very significant selling point, when compared to utility-scale photovoltaics. As we increase our reliance on intermittent renewable energy sources such as wind and photovoltaics, technologies that can ensure a well-modulated flow of power to the grid will grow in value.

CHAPTER EIGHT

Cradle to Grave

SOLAR ENERGY MAY BE nearly limitless, but capturing enough of it to edge out more harmful, non-renewable energy technologies is a formidable undertaking. As Sempra's Joe Rowling rightly observed when we stood amid his company's 2.4 million PV panels at the Copper Mountain solar complex in Boulder City, our very survival on earth depends on the broad and relatively benign diffusion of the sun's rays. Yet, because the sun's energy is so dispersed, we will need to cover lots of ground, roof space, and car parks with solar collectors if we are to make a serious dent in our reliance on fossil and nuclear fuels.

Today we get less than half-a-percent of our electricity from the sun, using tens of millions of PV panels to achieve this still-modest clean energy gain. Looking to a future when solar power makes its way through and beyond single-digit percentages, we can expect to see the number of solar panels deployed in America rising into the billions. Manufacturing all this hardware will require lots of energy and plenty of raw materials, and managing it responsibly at the end of its productive life will demand that we open new doors to recycling and disposal.

To understand what is involved in making solar panels, I decided to visit the SolarWorld factory in Hillsboro, Oregon, a few miles outside Portland. Not only is SolarWorld one of the few PV panel factories operating in America, but at the time of my visit it was the only US plant that carried out all phases of production at a single site.[1] Ben Santarris, the company's communications director, likes to call it a "soup-to-nuts" operation.

Ben greets me in the lobby and gives me a brisk summary of the company's history. SolarWorld was formed as a German start-up in 1998, but its acquired American lineage can be traced back to a

California-based firm called Solar Technology International, founded in 1975. When SolarWorld purchased the firm in 2006 from a successor company, Shell Solar, its manufacturing base was still in Ventura County. The move to Oregon happened two years later, when SolarWorld opened its new plant in a refitted microchip factory in Hillsboro. By 2010 one thousand people were working at the site.

Ben mentions a few of the more noteworthy places where Solar-World's PV panels have been installed. There's the Papal Audience Hall at the Vatican, several buildings at the Pearl Harbor naval complex, three installations at Yosemite National Park, and one of America's largest residential solar arrays at the hilltop ranch of former *Dallas* TV star Larry Hagman. Less notably, the Talmadge Road warehouse I visited in New Jersey has 17,745 SolarWorld panels on its roof, and our home in Massachusetts has 23 of them supplying most of our household power needs.

Handing me a pair of protective eyeglasses, Ben leads me from one section of SolarWorld's half-mile-long production line to the next, carefully describing each step in a precise, highly mechanized process that begins with the fabrication of silicon ingots and culminates in the framing of complete solar panels, often called "modules" in the solar trade. He points to a cluster of thick-walled vats filled with irregularly sized chunks of a silvery-gray mineral. "Silicon, as you probably know, is the most common hard element on the planet. It's literally under our feet." The raw material may be plentiful, but it takes lots of energy to convert it into long rods of polysilicon. The polysilicon, broken into large pieces, is what I see inside the vats. "I don't like to call them 'rocks' because that implies they exist in nature," he explains. Each vat is actually a crucible made of quartz, designed to withstand multiple days of intense heat without contaminating the molten silicon.[2]

We enter a high-ceilinged room filled with long rows of tall cylindrical furnaces, slim at the top, broader below. "So this is where it all starts," Ben declares. A crucible containing 250 pounds of polysilicon is loaded into each furnace's lower chamber, and a small amount of boron is added as a dopant, giving the crystal a positive electric orientation. The furnace is then heated to about 2,500°F—hot enough to melt the silicon. Walking over to one of the furnaces, we peer very briefly through a small orange-tinted window. Even when filtered, the light emanating from this superheated mass of silicon is intense. I understand why Ben calls it white heat.

On a computer screen, we can see a magnified photo image of what's happening inside the furnace. Suspended above the crucible is a slowly rotating, slender cable. At the lower tip of the cable, barely brushing the surface of the molten silicon, a crystal is beginning to form. "One . . . two . . . three . . . four." Ben counts the silicon's facets as they rotate in front of us on the computer screen. He likens the germinating crystal seed to a snowflake. Ever so slowly, it will grow into a smoothly rounded crystal eight inches in diameter and about 5.5 feet long. Once fully formed, the crystal has to cool down enough to be safely removed from the furnace. The whole process takes three-and-a-half days.

In a corridor outside the furnace room, Ben shows me a finished crystal, still too hot to touch, lying cradled beneath a large industrial fan. Once it reaches room temperature, its tapered top and tail will be sliced off, forming a perfect cylinder or "ingot." These ingots are the building blocks for "wafers"—the building blocks for photovoltaic cells.

Cutting ingots into six-inch-square discs that are 180 microns thick—about equal to the width of a human hair—is an exacting job. First, the rounded outer portions of the ingots are trimmed to create rectilinear blocks, each about the size of a long loaf of bread. (The trimmings, amounting to almost half the ingot's original volume, can be reused to make future ingots.) Then, like a high-tech bread slicer, a wafer saw cuts each silicon block into hundreds of ultrathin discs. It does so by running a single strand of wire, 400 miles long, along hundreds of minutely close, parallel paths. "That's about the distance from Portland to Missoula, Montana," Ben says of the continuous wire strand. The actual cutting isn't done by the wire, but by an abrasive compound that the wire picks up when it runs through a slurry of glycol and silicon carbide. Unlike a bread slicer, the wafer saw takes about six hours to grind its way through a single silicon block, and by the time all the cutting is done, half of the silicon has been turned to sawdust. Though much of this silicon waste can be separated from the remaining slurry via centrifuge, recovering it is a costly process and isn't always done.

Next we move to the phase where wafers are turned into photovoltaic cells. This particular step requires clean-room air quality: no more than a thousand particles per cubic meter—a minute fraction of the particles normally found in outdoor ambient air. To prepare, Ben and I suit up in thin mesh garments, from hairnets and full-body suits down to hygienic booties. We enter a room where long cartridges of wafers, vaguely resembling CD storage racks, travel robotically from one chemical bath to the next. One bath cleans the wafers, another roughs up their surfaces, creating microscopic pyramids that expand the area exposed to the sun's energy. Then one side of each wafer is treated with phosphorus gas under high heat to create a negative electrical orientation, assuring a consistent flow of photons from the cell's positively oriented reverse side. After these and other chemical treatments, the wafers are screen-printed with a matrix of fine metal strands and broader collector strips, or busbars, which will channel the flow of photons moving out of the cell.

To witness the final stages of production, Ben and I walk to a second building where individual cells are assembled into completed panels. We climb up onto a suspended walkway and peer down at a bustling, highly automated world. In the first assembly stage, strings of ten PV cells are soldered together, and six strings are then laid out on a rectangular pane of glass covered by a milky sheet of ethyl vinyl acetate (EVA), a laminating resin. After the cells are in place, they are covered by a second EVA layer, followed by a durable polymer backsheet. Heat is then applied to melt the EVA, fusing all those layers into a vacuum-sealed, laminated sandwich. Finally, an aluminum frame is placed around the sandwich, an electrical junction box is attached to the backsheet, and the panel is complete. All of this is done with rapid-fire robotic precision, and with almost no direct human contact.

As he leads me back to the reception area, Ben reflects on the contribution his company, and the solar industry generally, is making to the world of sustainable energy. Some years ago, he traveled to West Virginia coal country, where he witnessed one of the biggest rail yards in the nation. One coal train after another rolled through, each more than a hundred cars long, each carrying rock that once underlay green ridgelines in an area now torn apart by mountaintop removal mining. "It's great because it allowed for industrial production," he concedes about coal. "But it's outlived its day. It no longer makes sense." He contrasts this outmoded fuel source with solar energy. "Roughly half our power generation is based on taking rock and burning it up at a high heat, never to see it again, ever. And then that rock is gone, and it only kept your light on for an instant. . . . You can take the same amount of ore and turn it into a silicon wafer that generates electricity for twenty-five years or more."[3]

—m—

While many solar manufacturers, including SolarWorld, warranty their products for twenty-five years, those who study the life cycle of solar power systems generally assume that panels put in service today will continue producing substantial amounts of electricity well beyond that time frame. They are unlikely to maintain their full rated output as the years go by, however. In general, it has been found that solar panels and solar systems lose about 0.5 percent of their original productive capacity per year of operation.[4] Even at that degradation rate, we can count on solar power plants to generate far more energy than they consume over the course of their productive lives. By contrast, a power plant that relies on a carbon-based fuel like coal, gas, or oil creates an energy deficit from the moment its fuel is extracted from the earth, and that deficit only deepens as the plant advances through the years. Every kilowatt-hour of power is dependent on the burning of a polluting, climate-changing, non-renewable energy source.

Solar power follows a very different trajectory. Numerous studies have found solar power systems to recoup their energy inputs in a very small fraction of the typical power plant's operating life. One study has estimated that it takes 1.07 to 1.26 years to recover the energy invested in manufacturing silicon crystal-based PV panels. An additional 0.25 years is required to recoup the energy used making other components such as mounting systems, cabling, and inverters.[5] Another study, focusing on commercial rooftop PV systems, found the energy payback period for various types of PV installations to range from 0.68 to 1.96 years.[6] Panels that use monocrystalline wafer cells, like those I saw being made at SolarWorld, have the longest energy payback time because of their high up-front energy demands, but they generally end up producing more power over the course of their useful lives than other PV technologies.

The quickest energy payback is achieved by thin-film solar panels, which sidestep the energy-intensive demands of growing silicon ingots and transforming them into PV cells. Instead, thin-film manufacturers use one of various chemical vapor deposition methods to apply small quantities of a semiconducting compound to a substrate made of glass or, sometimes, stainless steel.[7] Cadmium telluride, the most commonly used thin-film semiconductor, has the advantage of capturing a relatively broad spectrum of the sun's rays. Its toxicity is a major downside, however. Copper indium gallium selenide and amorphous silicon are other semiconductors used in thin-film manufacture. Fortunately, the amount of semiconductor material used in thin-film production adds up to a tiny fraction of a thin-film panel's overall weight—much less than the amount of silicon used in crystalline cells.[8] Thin-film panels are cheaper and less energy-intensive to produce, but they are much less efficient than crystalline silicon in capturing the sun's energy.

A key PV performance measure is "conversion efficiency," or the percentage of the sun's total irradiance that a solar cell is able to convert to electricity. Conversion efficiency is generally gauged as the amount of electricity produced when a cell at 25°C (77°F) is exposed to 1,000 watts of sunlight per square meter. Because it is difficult to obtain uniform measurements outdoors, most testing is done in a laboratory setting, where quick flashes of light mimic the sun's rays. Under standard test conditions, a cell that is 17 percent efficient produces 170 watts per square meter; a cell that is 18 percent efficient generates

180 watts per square meter, and so on. It is important to note, though, that the power delivered by a solar panel is a few percentage points below the conversion efficiency of its solar cells, due to optical losses from the glass and EVA encapsulant and power losses that occur in the panel circuitry. Additional power is lost in converting the panel's DC current to AC current.

Also worth noting is the somewhat counter-intuitive fact that PV modules perform better in cooler weather. A PV cell exposed to direct sunlight when the outdoor temperature is 77°F will become much hotter than the ambient air, causing its output to drop off quite substantially. This preference for colder operating temperatures is one of the reasons why PV systems work so well in cold but relatively sunny states like Massachusetts and can even produce lots of power in frigid places like Alaska.[9]

Crystalline silicon cells on the market today have a mean conversion efficiency in the 17 to 19 percent range, but one American company—SunPower—leads the pack with its X-Series solar panel.[10] The cells in this top-rated panel have a conversion efficiency of 24 percent, and the panels themselves operate at 21.5 percent efficiency.[11] Though SunPower does most of its manufacturing in Malaysia and the Philippines, it is headquartered in San Jose and is a major developer of US utility-scale projects.[12]

The conversion efficiency of thin-film panels is generally lower than that of crystalline silicon panels, though thin-film efficiency ratings are on the rise. First Solar, the leading US company in the thin-film business, reached 13.9 percent efficiency on its top-performing production line by the end of 2013, and the company has set 19.5 percent efficiency as its target for 2017.[13]

—◊◊◊—

Various innovations are now being tried out as ways to ramp up the efficiency of solar photovoltaics. One of them is the bifacial module, which—as its name implies—has solar cells on both of its surfaces. The purpose of this design is to capture not just the direct sunlight beaming down on a panel's upper surface, but also the indirect light that reaches the underside of the panel.[14] Given the lower amounts of energy that can be drawn from indirect sunlight, doubling the number of solar cells translates into far less than a doubling of the electricity generated. Adding a second photovoltaic surface may therefore be hard to justify economically until solar cells drop significantly in price.

While it's not clear what the upper limit to solar cell productivity might be, researchers are now exploring ways to broaden the spectrum of solar energy that can be converted to electric power. Since different semiconducting materials respond to different wavelengths or "band gaps" of light, it has been found that stacked layers of diverse semiconductors can produce greater amounts of electricity than cells that rely on a single semiconductor. The top layer might capture blue light; a middle layer might be sensitive to green; a bottom layer might respond well to red.[15] The National Renewable Energy Laboratory has achieved 31.1 percent conversion efficiency with a two-junction cell, and Sharp Corporation of Japan has hit 44.4 percent efficiency with a three-junction cell.[16]

Nanotechnology may offer other pathways to greater cell efficiency. Still in the early stages of lab experimentation, nanocrystals—also called quantum dots—are being used to capture a broader spectrum of sunlight. In theory, these microscopic semiconductors can access more than 66 percent of the solar spectrum. In practice, though, it remains to be seen how much of that light can be economically converted to electric power.[17]

Because of their much higher cost, new technologies like multijunction solar cells are used almost exclusively in ultra-high-value applications like the powering of satellites, and cells based on nanotechnology have not been commercialized at all. Before these innovations can enter the broader marketplace, they will have to overcome consumers' preference for technologies that are known, trusted, and much more affordable. As one expert puts it: "The[se] exotic technologies will need to battle an entrenched, low-cost incumbent."[18] For the near term, incremental improvements in crystalline and thin-film conversion efficiency may be our brightest hopes.

—⚇—

A solar panel may perform extremely well when it's brand-new, but how long will it maintain that level of performance, and what are the major factors affecting its durability? These are questions that Dr. Geoffrey Kinsey and his colleagues at the Fraunhofer Center for Sustainable Energy Systems grapple with every day. Equipped with advanced degrees in physics and solid state electronics, Kinsey was hired to head up Fraunhofer's PV division after holding a number of senior research positions in the solar industry.

Fraunhofer's mission is to go way beyond traditional product rating labs, which look at PV panels' initial performance but do little to ascertain how well and how long they will hold up in the field. Kinsey says this brand of "infant mortality" testing is a very poor match for solar technology. "Most of our industry isn't built around things that last around twenty-five years," he says. Consumer electronics have an expected lifespan of three to seven years. Military hardware, including satellites in orbit, should last about twice as long. "Think about the enormous investments that are going into launching satellites, and they're really only designed to last fifteen years!"

Christian Honeker, an MIT-trained materials scientist, interjects that, in testing the durability of PV panels, it's important to look at the individual materials used, but equally vital to assess how different materials interact with one another. "That's where the complexity starts," he says. "You combine different materials, each of which you might know quite well, but that combination behaves differently." This is especially important in a fiercely competitive industry where manufacturers are tempted to cut corners. "There's been a huge pressure on cost," he notes, "to the detriment of quality."

Honeker points to the EVA encapsulant, used to seal off the inner workings of solar panels from the elements, as one component that is particularly vulnerable to damage resulting from the use of cheaper materials. "If you don't choose your additives correctly, the encapsulant might yellow. It might become brittle. It might delaminate." He calls delamination "a crack in the armor" that can lead to moisture penetration, a problem that is heightened in places like New England, where freeze-thaw cycles can widen fissures and accelerate degradation. When moisture penetrates, solder bonds can corrode, and bubbles forming on cell surfaces can diffract sunlight and reduce cell productivity. Delamination can also allow air to enter, causing the silver grid lines that collect the electrons to oxidize. The net result: "a gentle loss of power" that, over time, can substantially damage performance.

Fraunhofer has devised a set of grueling ordeals to gauge PV panel durability. In one test, panels are exposed to elevated heat and humidity over prolonged periods to see if encapsulants lose their adhesion. Rapid cycling between extreme heat and cold is also performed in a machine with a trade name worthy of Orwell: the Thermotron. Honeker prefers to call it "the torture chamber." In yet another test, panels are subjected to heavy physical loads and freezing temperatures, suggestive of snowfall.

Simulating the life cycle of a solar panel through accelerated testing is as much art as science, Kinsey admits. "It's a really tough challenge scientifically to say X hours in a chamber equal Y years outside," he says. For that reason, multiyear tests of panel degradation are also conducted under actual weather conditions at Fraunhofer's outdoor testing facility in Albuquerque, New Mexico. Fraunhofer's scientists gather the best data they can from these multiple sources, and they use that data to help solar manufacturers improve the durability of their products.[19]

—〰—

Unless there is a major breakthrough in the way PV systems perform, the typical solar panel will reach the end of its productive life span in twenty-five to thirty-five years. At that time, we may be replacing older panels with newer, more efficient ones, but we will not be able to run away from the mountainous waste burden created by our first truly big wave of solar power generation. This waste dilemma may not seem pressing today, but policy makers, industry leaders, and citizens' groups have already begun to wrestle with key questions: Should PV panels be considered hazardous waste, and if so, do existing hazardous waste laws adequately govern their safe and proper management? To what degree do current rules on electronic waste management provide useful guidance? And how should the responsibility for managing PV waste be shared among panel manufacturers, vendors, and PV project owners?

Not surprisingly, California has stepped out ahead of other states in seeking to establish clear ground rules for managing PV waste, just as it is leading the nation in the use of solar energy.[20] In July 2010 the state's Department of Toxic Substances Control sought public comment on its proposed steps to ensure the safe but efficient management of discarded PV panels by designating them as "universal waste," a middle-ground category requiring measures more stringent than ordinary solid waste but less rigorous than hazardous wastes covered by the federal Resource Conservation and Recovery Act (RCRA).[21] Under the proposed regulations, discarded panels would be delivered to authorized reclamation facilities only, where their reusable contents would be recycled and their hazardous constituents recovered.[22]

The California regulations have yet to be finalized, and some other states are relying on existing laws governing electronic waste. Depending on the specific safeguards built into these laws, the protections they afford may or may not be a good match for PV waste. In any case, absent a consistent set of nationwide rules, solar waste handlers will inevitably shop around for the cheapest and easiest places to unload their PV hardware. States with the most lenient waste provisions risk becoming the dumping grounds for PV waste from states with stronger safeguards.

The risk of cross-border dumping takes on a global dimension in the absence of any clear prohibition against shipping PV waste outside the United States. More than two decades have passed since the Basel Convention on the Transboundary Movements of Hazardous Wastes and Their Disposal entered into force in 1992. This global treaty regulates the transfer of hazardous wastes from the territory of one treaty member to another, and it bars the shipment of wastes to non-member nations. The European Union plus 180 individual nations have ratified the convention, but the United States has refused to do so. Solar waste handlers will therefore face few obstacles if they choose to shift the PV waste burden to overseas locations lacking the infrastructure or regulatory safeguards to handle them properly.

Despite the absence of a unified domestic or international framework for managing PV waste, a few manufacturers operating in the United States have voluntarily launched their own take-back programs. First Solar, for example, has operated recycling facilities at its US and overseas factories since 2006, and claims to have recycled 48,000 metric tons of its thin-film hardware in the intervening years. Through a multistage process of shredding, crushing, rinsing, and separation of solids from liquids, the company is able to recycle about 90 percent of a panel's total weight, mostly made up of glass. More significantly, it recycles 95 percent of the cadmium-telluride semiconductor material used in its panels—especially important given the known toxicity of this compound.[23] (It's worth noting that, while crystalline panels use small amounts of silver and other metals, they do not pose the same contamination risk as some of their thin-film counterparts.)

First Solar's recycling has been a technical success, but the company has stumbled in its effort to sustain the world's first pre-funded panel collection and recycling program. For a number of years, it charged customers a per-panel fee at the time of purchase and placed the collected funds in a restricted investment account dedicated to collection and recycling. This prepayment approach was intended to guarantee that PV waste would be handled in an environmentally responsible manner even if the company itself became insolvent. Eventually, though, First Solar found it too costly to maintain this program in the face of growing price competition from Asian PV producers. "Commercial customers . . . don't want to pay upfront for recycling services that will be needed twenty-five years from now," company vice president Alex Heard explained when First Solar gave up its pre-funded program in 2013. He promised, however, that the company would honor its prior commitments to customers and would continue to invest in recycling its own hardware.[24]

Sheila Davis, executive director of the nonprofit Silicon Valley Toxics Coalition, understands why companies like First Solar would abandon prepaid recycling. "The industry is so competitive that it becomes a disadvantage for you to take back your panels and absorb the cost of recycling them when your competitor does not, and when the margins are so narrow," she says. Yet, as a longtime advocate for responsible environmental management in the electronics industry, she laments how readily companies put their environmental safeguards on the chopping block. "Companies will adopt environmental policies, but when times get tough and they need to cut corners, that's sometimes the first thing to be cut."

Davis is convinced that clear and consistent rules need to be adopted to bring the solar industry as a whole to a new level of environmental and social responsibility. She takes some heart in the partial success of her group and its allies in pressing for e-waste recycling laws, now in effect in more than two dozen states. "I've been working on electronics for almost fifteen years," she says. "There has been steady change getting landfill bans adopted, passing legislation, and getting companies to take back their products—from a starting point when they wouldn't even talk to us." Unlike the electronics industry, which for many years turned its back on groundwater pollution and worker safety hazards, Davis sees much greater motivation among solar industry leaders. "Electronics companies never were really thinking about their environmental impact until we called them out on it," she remarks. "Solar companies want to maintain their green branding."[25]

As we begin laying the groundwork for solar waste management here in America, we can learn from the European Union's recent progress in creating a new norm of extended producer responsibility for PV waste. Under the EU's Waste Electrical and Electronic Equipment Directive, sellers of electronic equipment must accept, free of charge, discarded end-of-life products brought in by their customers. In 2012 the directive was extended to include PV panels, requiring EU member states to adopt and implement national laws in accordance with this mandate. By 2015, 70 percent of discarded PV panels are to be recycled, and by 2018 that target rises to 80 percent of panels entering the waste stream.[26]

Even before PV recycling became mandatory in Europe, a non-profit organization called PV Cycle set about building a region-wide PV recycling regime. Today PV Cycle is Europe's leading take-back and recycling service for PV technology, with over three hundred collection points where households and other small-scale PV users can discard their panels. For larger quantities of PV waste, there is a scheduled pickup service. Collected panels are taken to authorized recycling centers—all within Europe. Though the administration of take-back schemes varies from country to country, PV Cycle's services are supported through fees paid by PV manufacturers and vendors, and are based on each company's sales within a given national market.[27]

The route to PV recycling in America is far less clearly mapped. In some cases, the value and sheer volume of recyclable materials may spawn new or retooled businesses eager to reclaim at least some of these resources. For example, the metals used in many frames and mounting systems are easy to recover and have real value as recycled commodities.[28] Other materials, however, may be too costly to extract or have too little value to attract recycling entrepreneurs. Glass—constituting about 80 percent of a typical panel's weight—can be crushed, cleaned of chemical residues, and recycled, but its market value alone may not justify the investment. Semiconductor materials can be recovered, but at a cost that may deter private investment unless their recycling is required.

Dustin Mulvaney, a professor at San Jose State University who advises the Silicon Valley Toxics Coalition, is convinced that voluntary recycling will work no better in America than it worked in Europe before PV recycling became mandatory in 2012. He points to the "free rider" problem, whereby the voluntary participants in a recycling program lose market share to non-participants. "Companies will say, 'Why should I be spending my additional money for the end-of-life when, in fact, there's no one requiring me or forcing me to do so.'"

Mandatory, sustainable management of PV waste will place administrative demands on industry players as well as the government agencies overseeing them; it will also come at a financial cost. These burdens, in Mulvaney's view, are a fitting quid pro quo. "We subsidize this product with public dollars far more than any other electronic products," he says. "Maybe we *should* be holding it to a higher standard."[29]

—✺—

Though we may be slow in facing up to the PV waste management issue, America's growing investment in solar energy has already begun to reap notable environmental benefits. With every new kilowatt of installed PV capacity, we are increasing our reliance on an inexhaustible energy resource whose carbon footprint is a fraction of the size of its fossil fuel counterparts. According to the Intergovernmental Panel on Climate Change (IPCC), electricity generation from solar PV emits about 46 grams of carbon dioxide-equivalent per kilowatt-hour—and that includes the full life-cycle emissions from this power source: the manufacture of PV panels and related equipment, the operation and maintenance of power plants, and the dismantling of PV arrays at the end of their useful lives. Concentrating solar power has an even smaller life-cycle carbon footprint of 22 grams of CO_2-equivalent per kilowatt-hour. By contrast, the IPCC rates life-cycle carbon emissions from coal plants at 1001 grams per kilowatt-hour, and carbon emissions from natural gas-fired plants at 469 grams per kilowatt-hour.[30]

Since PV power plants consume very little off-site electricity during their operational lifetime, the carbon footprint of a PV installation depends largely on the amount of energy consumed and the type of fuel used in manufacturing PV hardware. A solar panel factory in China that relies on coal-generated electricity creates a much bigger carbon footprint (and emits much higher levels of other dangerous air pollutants) than a manufacturing plant in the United States or Europe that relies on a mix of renewable and non-renewable fuels.[31] The fuel used to transport panels to the United States from China and other solar-exporting Asian nations only adds to the pollution associated with imported PV technology.

To catalyze a shift to cleaner solar power, the US solar industry and consumer groups could perform a valuable public service by calling on all solar manufacturers—importers as well as producers—to report on the life-cycle carbon footprint of their products. A similar feedback loop could help reduce the conventional air pollutants—sulfur dioxide, nitrogen oxides, and particulates—associated with solar manufacturing.

Finally, we can look to improved PV conversion efficiency, slow though it may be in coming, as an added means of minimizing the environmental impacts of solar technology. Increasing conversion efficiency will not only make better use of the energy and materials associated with PV manufacture; it will also reduce the land area and roof space required to attain a given solar output. As we scale up our reliance on solar power, finding ways to minimize conflicts over the use of our built and open spaces will be essential to the development of a robust solar energy sector.

Building a Robust Solar Economy

FOR MILLIONS OF AMERICANS, "Solyndra" became synonymous with wasteful government spending during the heated 2012 electoral season. In his bid for the presidency, Mitt Romney used this solar panel maker's bankruptcy to cast a long shadow over the Obama administration's support for renewable energy. His own hyperbolic rhetoric was joined by that of numerous members of Congress—Republicans for the most part—who saw easy prey in Solyndra's shutdown of its Silicon Valley manufacturing plant less than two years after the Department of Energy approved a $535 million loan guarantee to the company.

With Solyndra's glass-walled headquarters and its large red "FOR SALE" banner as a backdrop, Romney unabashedly beat the drum of unfettered capitalism as he spoke to a small group of reporters. "The president fails to understand the basic nature of free enterprise in America," he said, his forehead gleaming in the sun as he stood in an immaculately pressed shirt and sky-blue tie. "He thinks that government-dominated decisions like this make America stronger. They make us weaker."

This was strange criticism coming from a man who, as governor of my home state, Massachusetts, had approved millions in government support for another US panel manufacturer that met a fate very similar to Solyndra's. In March 2011, after just a few years of operation, Evergreen Solar closed its factory in Devens, Massachusetts, laying off eight hundred workers. Its expressed intent at the time was to shift production to China, but the company declared bankruptcy a few months later.[1]

But Romney was on a roll that sunny California morning. He

branded Solyndra's building "the Taj Mahal of corporate headquarters," pointing to the LCD temperature readouts in company showers
and the Disney tunes that were piped into factory areas.[2] These minor amenities were, of course, not why Solyndra failed. The company
had banked on the market being ripe for an innovative solar panel
whose cylindrical cell design made minimal use of polycrystalline
silicon, considered at the time to be one of the biggest cost factors
in PV manufacture. Though sparing of silicon, Solyndra's tubular
cells turned out to be more expensive to produce than conventional
crystalline silicon cells. So long as the price of polysilicon remained
high, Solyndra's prospects looked good, but when polysilicon prices
plunged in 2009, the company's fate darkened. Despite a successful
push to cut production costs in half, its panels still couldn't compete
with the flood of cheaper Asian-made panels that were making their
way to the global marketplace.[3] Prices for solar arrays plummeted 42
percent during the first eight months of 2011 alone.[4] On August 31 of
that year, Solyndra announced its insolvency.

At the same time that Solyndra's fate was being sealed, other US
solar manufacturers were becoming increasingly alarmed by the sharp
decline in solar PV panel prices and China's rapidly growing share
of the market. SolarWorld, with its newly built $600 million factory
in Hillsboro, Oregon, had the most to lose as the biggest producer
of solar panels in America. Its concerns were reflected in a series of
ads featuring Larry Hagman of *Dallas* fame. At the time, Hagman's
California ranch boasted a 94-kilowatt solar array that heated his
swimming pool, pumped water for his water fountains, and cooled
his 25,000-square-foot home in Ojai, California.[5]

Though no energy miser, the TV impersonator of oil tycoon J. R.
Ewing had become a true believer in solar energy. He signaled that
devotion, along with his disdain for cheap Asian technology, in an
ad still viewable on YouTube. As the ad opens, he is approached by
a small boy wearing a Stetson hat just like the aging actor's. The boy
holds out a battered toy helicopter. "Grandpa, it's broken," he whines.
Hagman peers at the copter's belly to figure out where it was made.
"Oh well," he mutters. "It's always like that with these things." The

Larry Hagman, former *Dallas* TV star, promoting
SolarWorld panels. (Photo credit: SolarWorld Americas)

word "China" is not mentioned. "What about *your* toy?" the boy then
asks, causing a smile to flash across Hagman's otherwise stern face.
"Well, *my* solar panels are made in Germany. They're good for twenty-
five years—guaranteed." He hands the helicopter back to the boy and
snaps: "Take that piece of sh-t and throw it in the can, OK?"[6]

SolarWorld touted its German-engineered products, but it signaled a big-time commitment to American manufacturing when it
opened its 750,000-square-foot factory in Oregon. Its grip on the
American market started to loosen, however, soon after solar panels
started rolling off the Hillsboro assembly line. As SolarWorld's man-
agers viewed it, lower-priced but inferior-quality Chinese solar panels
were clearly to blame. In 2008 China accounted for 8.6 percent of all
crystalline silicon PV cells and panels imported to America. By 2010
its share of US imports had risen to 20 percent. The following year it
topped 45 percent.[7] This upward surge in Chinese imports roughly
mirrored worldwide trends in solar manufacture. In 2012 eight of the
top ten global solar manufacturers were Chinese, and 64 percent of

worldwide production came from factories in mainland China. The rest of Asia accounted for 21 percent of global manufacturing, while European factories supplied less than 4 percent of the market and only 3 percent came from the United States.[8]

What most alarmed SolarWorld executives was the drop in solar panel prices that accompanied China's rising market share. In 2007 the average global price for crystalline silicon-based PV panels hovered just below $4 per watt. By 2010 it had dipped under $2 per watt, and by 2011 it had descended to $1.28 per watt.[9] Railing against China's role in driving these prices so dramatically downward, SolarWorld's founder and CEO, Frank Asbeck, pointed to what he saw as the cause: "an authoritarian, non-market approach to business and industry, combined with a pattern of disregard for its intellectual-property and trade-law obligations to trading partners." As early as 2008, Asbeck noted the increasing presence of Chinese companies exhibiting at US solar trade shows. In 2010, when the United Steelworkers filed a complaint against Chinese trade practices in both the solar and wind energy sectors, he and his SolarWorld colleagues were still watching nervously. "We were grateful that someone was doing something—and not us," he noted, but conditions continued to worsen. "China was swamping and ultimately drowning US producers in an ever-rising tide of Chinese solar products," he lamented.[10]

In October 2011 SolarWorld decided to step into the fray, filing its own set of trade complaints with the US Department of Commerce and the US International Trade Commission. It focused these complaints on China's production of silicon-based PV cells, on the assumption that this stage of the manufacturing process was causing the greatest price distortions. Joining SolarWorld's petitions were five or six other solar companies that insisted on keeping their identities out of the public eye, fearing Chinese retaliation. These fears later appeared to be validated when, in May 2014, the US Department of Justice charged five Chinese military personnel with infiltrating confidential corporate databases at SolarWorld as well as two American steel companies that had been involved in trade disputes with China.[11]

To prevail in its trade claims, SolarWorld and its co-petitioners had to prove that Chinese solar manufacturers were selling, or "dumping," their products in the United States at artificially low prices, that the Chinese government was providing excessive subsidies to these companies in violation of relevant trade agreements, and that these actions were causing SolarWorld and its co-petitioners to lose sales or sell their products at lower, unprofitable prices.[12] The investigations bore out all these claims, leading the US government to impose tariffs of 31 percent, and sometimes higher, on Chinese panels containing Chinese-made cells. Chinese panel manufacturers were quick to circumvent these tariffs, however, substituting Taiwanese cells for those made in China but keeping all other production factors the same. In this way, they were able to avail themselves of what a leading solar trade journal called "a loophole large enough to steer a fleet of container ships through."[13]

SolarWorld, meanwhile, continued to see its sales and profits dwindle—so much so that, in July 2013 the company announced that it would soon lay off one hundred of its factory line workers and suspend production of silicon ingots and wafers at its factory in Oregon. Those layoffs continued an already steep slide in SolarWorld's Hillsboro workforce. From a peak of eleven hundred workers, only seven hundred were still employed at the factory in 2014.[14] Ben Santarris, my guide at the factory the previous year, placed the blame squarely on Chinese competitors. "Prices have continued to collapse as a result of Chinese dumping," he told the local press. "We need to be decisive in navigating this artificial crisis."[15]

In one more attempt to close the price gap, SolarWorld—going solo this time—filed a second set of complaints in December 2013, seeking to extend US tariffs to Chinese panels with cells produced in Taiwan or other third-party countries. The US government decided once again in SolarWorld's favor, slapping even higher tariffs on panels using third-party cells.[16] This came as a serious blow to Chinese manufacturers, about 70 percent of whom had shifted to rely on Taiwanese cells for their exports to the United States.[17]

By pursuing these punitive tariffs, SolarWorld not only placed itself at the center of one of the highest-profile US–China trade battles in recent decades; it also created a deep and bitter rift within the American solar industry. On one side was a small group of US solar manufacturers, joined by a contingent of US installers who had built their marketing around the use of American-made products. On the other side were Chinese solar companies as well as solar project developers, equipment suppliers, investors, and installers—all stakeholders who had good reason to fear that slapping tariffs on low-cost Chinese panels would drive up the cost of solar projects and send this new American energy sector into a slump, just as it was beginning to gain real momentum. SolarWorld and its allies formed one group, the Coalition for American Solar Manufacturing (CASM); those opposed to the trade sanctions launched their own initiative, the Coalition for Affordable Solar Energy (CASE). SolarWorld's Frank Asbeck dismissed CASE as "a cadre of Chinese manufacturers and downstream US businesses that had made quick riches on the dumped pricing of Chinese imports."[18] CASE president Jigar Shah, in turn, lambasted SolarWorld for its readiness to place the interests of a single enterprise above the greater good of a growing industry. "This is the type of economic opportunity that comes once in a generation," he wrote in a letter to President Barack Obama, "and it is now at risk due to the reckless self-interest of one German-owned solar panel manufacturer. . . ."[19]

The Solar Energy Industries Association (SEIA), America's leading solar industry trade group, has been caught in the middle. Along with representing manufacturers like SolarWorld, its membership includes US subsidiaries of Chinese solar companies, US companies that sell high-value factory machinery and manufacturing inputs like polysilicon to China, and solar developers who have benefited from the drop in panel prices. Despite the objections of some of its members, SEIA backed SolarWorld's original complaints and supported the creation of CASM.[20] As the case progressed, however, SEIA shifted from support, to equivocation, to outright opposition to the second round of petitions and its resulting penalties. When the Department of Commerce issued its provisional findings for the second round of tariffs in June 2014, SEIA president Rhone Resch stated: "These damaging tariffs will increase costs for U.S. solar consumers and, in turn, slow the adoption of solar within the United States."[21]

John Smirnow, SEIA's vice president for trade and competitiveness, has sought to bring about a shift from adversarial trade proceedings to a negotiated settlement—much as the European Union has achieved in its dealings with Chinese solar manufacturers.[22] Under the arrangement Smirnow envisions, all orders, investigations, and related proceedings would be terminated in exchange for Chinese producers paying into a fund that would help offset their price advantage over US products. The settlement would also address China's retaliatory trade measures, which have imposed tariffs on the import of US-manufactured polysilicon, a crucial feedstock for photovoltaic cell manufacturing in China. "More and more litigation doesn't help this industry," Smirnow says. "Our job is to try to find a path out of this that recognizes the interests of the entire solar industry rather than the interest of just a few companies."[23]

—⁂—

For a few years now, I've watched the warring factions stake out their widely divergent positions on US-China solar trade, and I've wondered: Should American panel manufacturers simply throw in the towel in this industry as in so many others, recognizing the marketing edge that lower-wage economies have in mass-producing so many of the goods we rely on for our comfort and convenience? Or is there a middle ground that defines a productive and profitable role for American enterprise, capitalizing on American technology innovation and benefiting from a higher level of manufacturing quality control than purchasers can count on when buying from a widely dispersed Asian supply chain?

When I posed these questions to SEIA's John Smirnow, he focused on branches of American solar manufacturing other than the production of panels. Polysilicon was his first example: America is a leading global producer of this basic feedstock for silicon-based PV cells. (Somewhat paradoxically, much of this high-value material is incorporated into Chinese solar panels that end up back in the United States, where they are sold to US solar customers.) Encapsulants (used to seal cell assemblies), electrical junction boxes, cables, and inverters are also produced in considerable quantities in the United States, as are the racking systems used on rooftop installations and the mounting systems for ground-mounted solar arrays. Smirnow says that about thirty thousand American workers are employed in solar manufacturing, but only a few thousand of them are directly involved in panel production.[24]

I also turned to Shayle Kann, a 29-year-old solar industry analyst at GTM Research. Kann and his colleagues bring rigor as well as enthusiasm to an industry that has become a magnet for smart, young, idealistic professionals. They share a sun-splashed, open-plan office above the Black Rose Irish pub in Boston's financial district.

As we begin our conversation, Kann rattles off what must be a well-practiced statement of his company's mission: "We are a full-service market analysis and consulting firm focused on the global transition of the electricity industry to a more renewable, sustainable, distributed, transactive market." I'm struck by how well these words reflect the transformative forces at play in solar power's development. Then, disarmingly, he adds: "We're a lot of young people who are excited about the prospect of solar and want to rethink how electricity works."

GTM Research has prepared what is probably the definitive analysis of the US–China solar trade dispute.[25] Kann shares some of that report's findings, and he offers his own candid views on the limits to tariff-seeking as a way to keep American solar manufacturers in business. The trade sanctions won by SolarWorld are "a blunt instrument," he feels. "It's in some ways a noble pursuit to try to make sure that we can continue to be a big player in the solar manufacturing business," he concedes before adding: "I just don't think this is the right way to go about it."[26] Kann cautions me not to assume that cheap labor is the key contributor to China's winning the solar price war. There are other Asian nations where labor is cheaper. What China has achieved is extraordinary scale, including a fully built-out domestic supply chain that meets most of the material needs of the country's solar manufacturing giants. Have government subsidies played a major role in strengthening that supply chain? Yes. Were those subsidies—the free or cheap land, free electricity, interest-free financing, and export subsidies—in violation of international trade laws? Yes, in Kann's view. But will tariffs succeed in making American manufacturers the primary suppliers of US solar panels? Not likely.

The most successful American solar panel manufacturers, Kann observes, are the handful of large, vertically integrated solar companies that develop, install, and sometimes finance solar power plants in addition to producing solar panels and other related solar hardware. SunPower, whose work I witnessed at the California Valley Solar Ranch, is a good example. Its panels, widely regarded as market leaders in efficiency and durability, are *not* made in the United States, even though the company is headquartered in San Jose, California. SunPower purchases polysilicon, ingots, wafers, solar cells, and various "balance of system" components like inverters from multiple sources around the globe, and it has its own solar panel assembly plants in France, Malaysia, Mexico, the Philippines, and South Africa.[27] In this important respect, it is a truly global corporation. Where the company shines is not in the stand-alone production and sale of solar panels, but rather in the delivery of high-output residential, commercial, and utility-scale solar power plants that incorporate its patented PV technology. SunPower is a solar system designer, developer, and installer every bit as much as it is a solar manufacturer. Through sophisticated sun-tracking equipment and the organization of its ground-based installations into super-efficient clusters of solar panels and inverters, called "power blocks," it has been able to offer customers a particularly high level of overall system performance.

SunPower CEO Tom Werner explains his company's strategy: "If you frame the competition as module competition, that's not what Western companies are good at," he says. "So what we do is reframe the competition. . . . Our customers, particularly the sophisticated customers, realize that the whole purpose of buying a solar system is to get economic renewable energy." The company's marketing approach was built around this broader aspiration. "We were able to say, 'Sure—you can buy a very low-priced Chinese module, though when you design it as a system and you install it, your cost of energy will actually be higher.'"[28]

As proof of SunPower's success in persuading customers to take this broader view, the company's communications director, Ingrid Ekstrom, backs her boss up with data on SunPower's American track record and global delivery of solar electricity. "SunPower is currently the leading manufacturer of residential systems . . . and the leading manufacturer and installer of commercial systems in the United States," she claims. Globally, by the end of 2013, it had deployed more than 2.5 gigawatts of solar capacity, with more than fifty thousand operating SunPower residential systems and more than one thousand commercial and utility-scale power plants.[29]

SunPower isn't alone among American-headquartered companies in relying on manufacturing supply lines outside the United States, nor is it unique in focusing on complete power systems rather than the sale of solar panels. Arizona-based First Solar, a global leader in thin-film solar panel production, does most of its manufacturing abroad and is a frequent rival to SunPower in lining up utility-scale solar power plants in America.[30] SunEdison is a Monsanto subsidiary that develops, finances, and operates solar projects in the United States in addition to producing solar wafers, cells, and panels. It has wafer factories in Texas and Missouri, but most of its production takes place overseas. The company is now weighing the feasibility of building a $6.4 billion crystalline PV panel manufacturing plant—one of the largest in the world—in Saudi Arabia.[31]

This general trend toward offshore manufacturing has its exceptions, however. SolarWorld's Hillsboro facility is still the largest solar panel plant in the United States, capable of producing half a gigawatt of solar power capacity per year. Soon, though, it may be matched or eclipsed by Silevo, a Silicon Valley–based company that has followed a path that GTM's Shayle Kann describes as one of the more promising niches for US solar manufacturing: the development of "early-stage, highly proprietary technologies" that are easily set apart from the conventional PV panels that are flooding the market today.[32] Silevo claims that its "tunneling junction" cell, combining aspects of thin-film and crystalline design, has already reached 22 percent conversion efficiency and has "proven headroom" to surpass the 24 percent threshold achieved by the pace-setting SunPower Series X cell.[33] In June 2014 SolarCity's visionary board chair, Elon Musk of Tesla fame, announced that the company had agreed to purchase Silevo for $200 million with the expectation that it would build a factory in New York capable of producing at least a gigawatt of solar power capacity per year. The plant, if built, will give SolarCity a guaranteed domestic source for its entire current solar inventory—more secure and, Musk believes, more economical than buying most of its inventory from China. Foreseeing tens of gigawatts of newly installed solar power per year in SolarCity's future, he regards the Silevo factory as a pilot plant for a much larger, dedicated US supply chain.[34]

—⁂—

The coming years will reveal the degree to which solar developers like SolarCity are able to reshape America's solar manufacturing marketplace by building their own dedicated supply lines on American soil. It's clear, though, that solar panels are by no means the only significant factor defining the price of a solar installation. When I look at the invoice that Sunlight Solar Energy prepared for my family's solar system, I see various costs listed separately: the panels themselves; the micro-inverters that convert each panel's DC power to AC; permit processing; and a catch-all category called "balance of systems" that includes hardware items such as rooftop mounting racks, electric meters, and emergency disconnect switches. Online data-monitoring software is also counted within "balance of systems." Stripping away all those add-ons, the nominal cost of panels used in residential solar arrays averages about one-fifth of the total installed price of a residential solar installation.[35]

Because "soft costs"—labor, marketing, permit processing, and the like—make up such a large percentage of the cost of installed solar systems, US solar developers are looking for ways to reduce those costs without sacrificing the quality of their installations. Germany is a source of inspiration because of its streamlined permitting, its uniform national solar incentives, its lower labor inputs, and its much lower customer acquisition costs. In 2011 the average cost of recruiting a US residential customer amounted to about $0.69 per watt. In Germany, acquiring new customers cost developers about $0.07 per watt.[36] This wide gap has been explained in part by the German public's much greater familiarity with, and receptivity to, solar energy. Back in 1999, Germany launched a determined effort to integrate solar power into households and businesses with its 100,000 Roofs Program. This government-led initiative garnered widespread media attention and broad public support. By the end of 2012, Germany had over four times as much installed solar capacity as the United States (32,400 megawatts compared to 7,800 megawatts), and Germany's solar power per capita, at 399 watts, was more than 16 times higher than America's meager 25 watts per capita.[37]

Selling solar systems is becoming easier as Americans grow more familiar with the technology and appreciate its economic as well as environmental benefits. With so many new customers, the competition among solar system installers and solar equipment vendors has intensified, and that competition, in turn, has helped bring down the installed price of solar systems. It's not just the big companies like SolarCity and SunEdison that are hustling for customers; it's your local roofer, electrician, architect, and general contractor who view this technology as a new source of business. According to the nonprofit Solar Foundation, solar jobs reached 173,807 by November 2014—up 21.8 percent from the previous year. That's nearly twenty times faster than the overall growth in US private sector employment during the same period. Employers anticipate a further rise in solar jobs in 2015 to more than 210,000 workers.[38]

As America's solar industry has matured and greater numbers of trained technicians, system designers, and financial planners have entered the workforce, it's no surprise that the price of PV solar systems has dropped, along with the price of panels. In the first quarter of

2012, SEIA and GTM Research calculated the average price quoted for a US residential solar array at $5.86 per watt. By the third quarter of 2014, SEIA and GTM had found that the actual installed price for completed residential installations had fallen to an average of $3.60 per watt. Non-residential systems—on-site installations at commercial, public, and nonprofit properties—showed a decline during the same period from $4.64 to $2.27 per watt, and utility-scale solar power plants had dropped from $2.90 to $1.88 per watt.[39]

What effect have these downward-sloping prices had on solar power's ability to compete with conventional fossil fuels and nuclear power? Lazard has taken a close look at the life-cycle economics of solar and other power generation sources, reflecting not only the cost of building and financing new power plants, but also the outlays required to operate and maintain them over a multiyear period approximating their expected operational lifespans. Lazard's methodology does not take into account some important externalities such as the differential contribution of each technology to health-impairing air pollution, climate-altering greenhouse gas emissions, or depletion of water resources. Monetizing these externalities would tilt the scales heavily toward renewable technologies like solar and wind. Also omitted are the costs of responsibly managing plant decommissioning—an important consideration for all electric facilities, including solar plants. Even with these gaps, Lazard's "levelized cost of energy" analysis offers a helpful tool for comparing the economics of our present and future technology choices.

For a utility-scale solar plant relying on crystalline silicon panels, Lazard estimates a levelized cost of 7.2 to 8.6 cents per kilowatt-hour—exactly the same as a PV facility using thin film solar panels. That's higher than the 6.5 cents per kilowatt-hour required by a new coal-fired power station that does nothing to reduce its carbon emissions, but it's much cheaper than building a coal plant that incorporates carbon capture and compression technology (15.1 cents per kilowatt-hour). Likewise, electricity from solar PV would be more expensive than power from a conventional combined-cycle gas plant (6.1 cents per kilowatt-hour), but it would be a bargain compared to building and operating a gas plant that is equipped to capture and compress carbon (12.7 cents per kilowatt-hour).

Thermal solar plants with storage capacity, like Crescent Dunes or Abengoa Solana, are much more expensive propositions. As mentioned earlier, Lazard prices these facilities at 11.8 to 13 cents per kilowatt-hour, just slightly below the levelized cost of a new nuclear power plant (12.4 to 13.2 cents per kilowatt-hour).[40]

Looking only at these numbers, it seems logical that US electric utilities would still want to invest in conventional gas and coal plants, so long as they aren't required to incorporate carbon sequestration technology. But solar has a number of special advantages that average life-cycle cost comparisons fail to reflect. Across much of the nation during the hot summer months, electricity demand spikes in the long afternoon hours, straining and often outstripping the generating capacity of baseload power plants—workhorses typically fueled by coal, gas, or nuclear power. Luckily, that's precisely when solar plants can deliver their greatest increments of power to the grid. Turning to solar at these times, utilities are able to reduce their reliance on "peak-shaving" power plants, reliant on natural gas or diesel, that are more expensive to run, less efficient in their use of fuel, and more polluting than their baseload fleets.

A further economic benefit to solar energy is its value as a hedge against price instability in the fossil fuel market. While natural gas–generated power may look like a cheaper investment than solar today, its future price is highly susceptible to geopolitical changes. Russia's recent aggression against Ukraine should be a warning in this regard. Though it possesses the world's largest known gas reserves, Russia's strained relations with its neighbors may cause European nations to turn elsewhere for their fuel supplies. Yet, given the political turmoil that plagues other leading gas-exporting nations (Algeria, Iran, Iraq, Libya, Qatar, and Venezuela, to name a few), global demand for US gas is very likely to rise substantially in the years ahead. Preparing for this opportunity, US gas producers are now seeking federal approval for a string of new liquid natural gas terminals along the Gulf of Mexico and a few more on the East and West Coasts.[41] A decade ago, before the fracking boom swept America, proponents of these terminals sought them as gateways for gas *imports* to the United States. Today the gas industry wants these same terminals to serve the *export* markets they are seeking to develop. As we hustle to meet our own increased demand as well as that of our global trading partners, US natural gas prices will certainly rise. The only questions are: how much and how soon? In March 2012 natural gas spot prices dropped as low as $2.04 per million Btu. In February 2014 they spiked as high as $5.92. By March 2015 they had settled back down to the three dollar range.[42] It's hard to predict where they might be in another ten to twenty years.

Coal is another wild card. If the US Congress ever summons the resolve to place a tax on carbon emissions, the cost of burning coal will rise, and if that tax remotely reflects the anticipated impacts of this fuel on the environment and public health, the cost of relying on coal will rise very substantially. Even short of a direct tax on carbon, pollution-control measures now being implemented by the Obama administration will substantially increase the cost of burning coal at existing power plants, and at new coal-fired plants if our utilities are brazen enough to build them.

With nuclear energy, the high price of new reactors is only one of the deterrents to expanding an industry with proven, devastating downsides. The Fukushima tragedy has highlighted the havoc that natural forces can wreak upon nuclear reactors and their surroundings. Chernobyl is an equally appalling reminder that catastrophic reactor accidents can occur under "ordinary" operating conditions, rendering entire regions uninhabitable for decades or even centuries. In an era of mounting global terrorism and cyber warfare, intentional assaults on nuclear facilities are a third specter hanging over the nuclear industry, only compounded by the risks of stolen nuclear materials becoming weapons in the hands of rogue states and non-state extremists. Nuclear power may produce fewer carbon emissions than fossil fuels, but is it worth proliferating this technology when safer, low-carbon alternatives are at hand?

—∿—

The future of US solar panel manufacturing may be uncertain, but if there's one takeaway lesson from the raging trade dispute with China, it's that tariffs are no panacea. Penalizing Chinese manufacturers for undercutting our own products may save a few hundred, or perhaps a few thousand American jobs, but the bigger challenge is developing the ways to build a robust US solar economy, strong enough to compete with the conventional energy sources that have dominated our electricity sector for too long. Where we can develop patents on new solar technology, we should do so. Where we can produce certain solar components more cheaply, or at higher quality, at home than abroad, we should expand those production lines. Where outsourcing mass-produced solar components can bring down the overall cost of solar power, we should invest in global supply chains, just as SunPower and First Solar have done.

Along with scrutinizing the policies that have supported solar manufacturers abroad, we need to cultivate and strengthen our own policies to make sure that solar energy can compete on a level playing field with other energy technologies here at home. The next chapter will look at some of the measures our own federal government and many states have taken to bring about a shift toward what Shayle Kann calls "a more renewable, sustainable, distributed, transactive" electricity market. Some fear the disruption that these policies may bring to the traditional, centralized delivery of power; others welcome the very same policies because of the fundamental changes they will bring. Both sides agree that if solar energy is implemented on a scale that is technically within our reach, it will fundamentally change the way we supply, store, and consume electric power.

CHAPTER TEN

Disrupting the Utility Status Quo

SOLAR POWER IS a newcomer to an industry dominated by players with a strong interest in preserving their primacy in the energy field. For decades, our coal, gas, and oil industries have been showered with government subsidies ranging from bargain-basement leases on mineral rights to tax breaks amounting to billions of dollars per year. Civilian nuclear power, spun off from untold billions in federally funded weapons research, has enjoyed ongoing government support in the form of research funding, tax breaks, a federally guaranteed cap on nuclear plant liability in the event of a catastrophic accident, and the right to continue operating despite the lack of any accepted solution to the nuclear waste quandary.

Along with these overt subsidies, America's fossil fuel–burning power plants have until recently been exempt from any meaningful restraints on their carbon emissions. The Obama administration's efforts to regulate CO_2 emissions from coal-fired power plants are an important first step, but we remain without a carbon tax, a nationwide emissions trading regime, or any other means of monetizing the environmental and public health costs of the greenhouse gas emissions produced when we generate electricity from carbon-based fuels. Recent government efforts to stimulate solar power development must be viewed against this backdrop of official neglect and government largesse supporting carbon-based fuels and nuclear power.[1]

Of all the government policies that have seeded the growth of American solar power, none has had a bigger impact than the 30 percent investment tax credit that the federal government has made available to solar investors large and small. Together with the precipitous drop in hardware costs, this incentive converted my own family from wait-and-see skeptics to enthusiastic buyers. To homeowners, businesses, and utility-scale solar developers, the tax credit has been an enormous boon.

Though Republicans have assailed the Obama administration for its pro-solar fiscal measures, the investment tax credit actually has its roots in the Energy Policy Act of 2005, signed into law by George W. Bush.[2] Initially the tax credit applied only to residential and commercial investments in solar and other renewable energy technologies, but Bush later allowed utilities to claim the credit. He also approved an extended timeline for eligibility, making the credit available to all qualifying projects placed in active service by the end of 2016.[3]

During the darkest days of the recession, Barack Obama took an additional step that made a huge difference to utility-scale solar developers as well as businesses installing their own solar systems. As part of his economic recovery package, the president allowed for the conversion of the investment tax credit to an outright Treasury grant.[4] At a time when potential solar developers lacked sufficient taxable income to avail themselves of the tax credit, this move made the difference between "go" and "no go" on a number of major solar projects. As of May 2014, grants disbursed by this program added up to nearly $6.6 billion, allowing 5.4 gigawatts of new solar power to be built—equivalent to supplying all the electricity needs of roughly nine hundred thousand American homes.[5] Though the application period for these grants ended in 2012, an additional $9 billion is available to developers that met this filing deadline so long as their projects enter into service by the end of 2016.[6]

Even with these sizeable government grants, the developers of utility-scale solar projects were in a bind during the cash-strapped recession years. This was a time when utility-scale solar power had yet to gain a foothold, so institutional lenders were reluctant to risk their constrained capital resources on a technology that they still viewed as financially unproven. The Department of Energy's clean energy loan guarantee program played a crucial role in filling this gap.

For most Americans, the politically charged furor over the failed loans to Solyndra and a few other clean energy companies obscured the broader, quite momentous accomplishments of the loan guarantee program. Obama administration officials pushed back against critics of the $34 billion program, asserting that investments in clean energy breakthroughs inevitably entail some level of risk. Why else, they reasoned, would government loan guarantees be needed? They also insisted that the program's losses, including the $535 million default by Solyndra and a failed $70 million loan to another panel manufacturer, Abound Solar, stayed well within the margin of losses that the Department of Energy had budgeted for.[7]

The federal loan guarantees played a particularly important role in launching utility-scale solar power plants. "The whole idea of utility-scale solar . . . did not exist before 2010," the program's executive director Peter Davidson reminded members of the Senate Energy and Natural Resources Committee when he was summoned to defend his program. During 2010–2011, the program ventured outside the comfort zone of private commercial lenders, extending credit to half a dozen photovoltaic facilities and another half-dozen concentrating solar power plants—all with generating capacities exceeding 100 megawatts. By July 2013, when Davidson reported to Congress, all of these projects were making good on their loans, creating a sufficiently safe investment environment for leading financial institutions—John Hancock, Bank of America, Citigroup, and others—to enter the field.[8] Indicative of the transformation that the loan guarantee program has helped catalyze over the past few years, utility-scale solar power has contributed more than half of all newly installed solar power capacity since 2012. (See fig. 6 in chapter 5.)

In its effort to help solar and other renewable energy technologies gain a foothold in utility-scale power generation, the federal government has gone beyond providing investment incentives and assurances. By opening up 285,000 acres of federal land for Solar Energy Zones, the Department of the Interior has sent a strong signal that Washington is serious about matching our nation's enormous land resources to the climate crisis that we face and the renewable energy opportunities that are upon us. Through its RE-Powering America's

Land Initiative, the Environmental Protection Agency has cleared a different pathway to solar energy development by mapping out the solar resources embedded in America's under-used and abandoned industrial lands. And the Department of Defense has begun looking closely at its own extensive land holdings as promising sites for large-scale solar energy development, motivated by its mandate to make its overall energy consumption 25 percent reliant on renewable energy by 2025.

Though less grand in scale, a number of other federal agency programs have begun to play a role in opening up solar energy's geographical and demographic horizons. With funds made available by the American Recovery and Reinvestment Act, the Department of Labor has given out grants to a number of nonprofits engaged in training unemployed minorities, veterans, and others to reenter the workforce with solar installation and other clean energy skills.[9] Solar arrays are also among the investments covered by the Department of Housing and Urban Development's $250 million Green Retrofit Program for Multifamily Housing.[10]

—◊—

As important as federal leadership has been to the mainstreaming of solar power, many of our states have been on the front lines of bringing solar technology to American households, farms, businesses, and electric utilities. Renewable electricity standards have been critical to creating a strong normative framework for deploying solar and other renewable energy technologies in twenty-nine states plus the District of Columbia, where they have been adopted. While California arguably leads the pack with its 33-percent-by-2020 mandate, Nevada has a standard calling for 25 percent of utility-supplied electricity to come from renewable sources by 2025, and as part of that broad requirement, solar energy must provide 6 percent of the electricity sold. New York's renewable electricity standard has an interesting twist: it requires "customer-sited" renewable energy—electricity generated at homes and businesses or renewable technologies such as solar water heaters that reduce electricity demand—to meet roughly 8.44 percent of the state's electricity needs, within a broader statewide commitment to achieving 30 percent reliance on renewable energy by 2015.[11]

To meet their renewable electricity mandates, many states offer a variety of economic incentives. Some, like the rebates offered to homeowners and businesses by state energy offices, are being scaled back as the economics of solar investments improve. Others, like the renewable energy certificates that solar system owners earn for the power they generate, fluctuate in value with utilities' demand for renewable electricity, as needed to keep pace with their statutory obligations.

Of all the state-based solar incentives, solar net metering is perhaps the most controversial. Net metering has been adopted, in one form or another, by forty-six states plus the District of Columbia.[12] In some of these jurisdictions, net-metered customers are credited for the full retail value of their surplus power, which is fed back into the grid and is available for other customers' use. Typically the amount of this surplus is limited by a cap on the size of net-metered PV systems, often expressed as a percentage of the customer's estimated annual electricity usage. In other states, fees and other pricing mechanisms are used to chip away at the economic gains of net-metered power. Utilities favoring these levies contend that solar customers should pay their fair share of the fixed costs for electric distribution and backup power; pro-solar advocates see utilities as penalizing net-metered customers for investing in a social good that actually saves utilities money. Nowhere has public debate over the shape of net metering policies been as bitter or high-profile as in Arizona.

Barry Goldwater Jr. may seem a strange champion of solar energy. After all, he's the son and namesake of a man who, as Lyndon Johnson's 1964 presidential challenger, declared that America would be a better country "if we could just saw off the Eastern Seaboard and let it float out to sea."[13] Barry Jr. established his own conservative credentials during fourteen years in Congress, but today he is Arizona's most ardent advocate for a power source that, until fairly recently, was widely viewed as the domain of liberal idealists.

As it turns out, defending solar power flows naturally from Goldwater's deeply held libertarian beliefs. Now the chair of Arizona's pro-solar coalition, Tell Utilities Solar Won't Be Killed (TUSK), he explains in a series of TV spots why he has become such a fierce defender of net metering. "Conservatives want—no, they demand— freedom of choice, whether it's health care, education, or even energy," Goldwater says in one of these ads. "We can't let solar energy be driven aside by monopolies who want to limit that freedom of choice. It's not the American way. It's not the conservative way." To drive the point home, a menacing wildlife scene flashes across the screen—of a large gorilla seizing and then tossing a much smaller gorilla to the ground. We then hear a deep male voice: "The 800-pound utility monopoly is trying to kill the independent solar industry in Arizona.... It wants to harm the little guy."[14]

Arizona Public Service (APS), the 800-pound gorilla in this ad, is the state's largest electric utility, serving most of the Phoenix metro area and several other cities. It became a magnet for solar advocates' indignation when it sought to impose new charges on residential solar customers, to make up for revenues lost because of lower electric- ity sales to those households. For APS, like other electricity provid- ers, many of the fixed costs of providing reliable power to its service territory are recovered through the fees it charges per kilowatt-hour. When power purchases plummet for a solar household, APS argues that ratepayers without solar arrays end up bearing a disproportionate share of the costs associated with building, upgrading, and maintain- ing transmission lines, distribution networks, and generating stations.

To get a better grasp of APS's reasoning, I met with Greg Ber- nosky, manager of state regulation at APS and formerly head of the company's renewable energy program. As recently as 2009, he told me, APS had just nine hundred solar customers. By 2013 that number had reached twenty-one thousand—out of a total of 1.1 million rate- payers. Solar installations were almost nowhere in sight as I scanned miles and miles of sunny sprawl from an upper-story window at the APS headquarters in downtown Phoenix, but APS is looking toward a future when customers with their own solar power systems will make up a much bigger share of the company's portfolio. This isn't just a fear; it's mandated by Arizona law, which requires 15 percent of retail electricity sales to be based on renewable energy by 2025. Of that renewable energy, at least 30 percent must come from "distributed resources" like solar arrays on residential, commercial, and public buildings, adding up to about 4.5 percent of total retail electricity. And of that 4.5 percent, half must come from residential installations.[15]

Bernosky tells me that by 2015, his company will get 11 to 12 percent of its power from renewable energy—way ahead of the 5 percent threshold mandated for that year by the state. While most of that power will come from large, utility-scale solar farms like the Solana parabolic-trough facility I visited in Gila Bend, he focuses on the problems posed by residential solar systems. He offers a hypothetical: a new residential development where most of the homes are equipped with solar. "There's still all the required infrastructure to get power delivery to everyone," he says. "We can't *not* provide the infrastructure assuming that solar will do the job." He has a point: solar arrays only produce power when the sun is out, so when householders cook their dinners, wash their dishes, run their washing machines, watch TV, and go online during the evening hours, APS has to supply all the power they need. And APS also has to be ready to absorb any surplus power that these same households feed into the grid during peak sunlight hours when their electricity use may be less than the output of their solar arrays. To qualify for net metering under Arizona law, a home solar installation must be sized to produce no more than 125 percent of the household's "total connected load"—the power consumed when all appliances, from light bulbs and computers to air conditioners and clothes dryers—are operating at their full rated capacity. Since normal use falls far short of this maximum, solar arrays can run up a sizeable surplus that APS and the state's other utilities are obligated to credit at the applicable retail rate.

In the summer of 2013, APS laid out a number of options for recouping more of its fixed costs from solar customers, including a surcharge per kilowatt-hour that would add at least $50 to the monthly electric bills of new solar customers. This triggered what Bernosky remembers as a "sound byte war," pitting APS against TUSK and other solar industry proponents who felt the utility was seeking to strangle the state's future prospects for distributed solar power.[16] Through its parent company, Pinnacle West, APS channeled $3.7 million into ads aired by the utility-supported Edison Electric Institute and two anti-tax groups called Prosper and 60 Plus. "We are in a political battle," APS spokesman Jim McDonald said to the local press. "We didn't ask for it. But we are not going to lie down and get our heads kicked in."[17]

In addition to their dueling ads, APS and its detractors deployed experts to catalog the costs and benefits of distributed solar. Drawing on a study it commissioned, APS estimated that each home solar array shifts $1,000 in costs per year to customers without solar.[18] A report commissioned by the Solar Energy Industries Association came to a very different conclusion. It found the economic benefits of residential arrays to exceed the costs to APS ratepayers by more than 50 percent, factoring in the avoided fuel, transmission, and distribution costs as well as the reduced need to build new conventional power plants.[19]

Weighing the different sides' widely divergent assessments, the Arizona Corporation Commission voted to impose a monthly fee of 70 cents per installed kilowatt on all new home solar installations, or about $5 per month for the average solar customer.[20] That was much less than APS had hoped to recover, but it did reflect an attempt to balance the state's interest in promoting distributed solar energy with its concerns about fairness to non-solar ratepayers. Commissioner Susan Bitter Smith stressed her state's important role in setting "the appropriate tenor for the conversation" in the many states where the debate about net metering has been heating up.[21]

That "conversation" has spread quickly. Colorado's biggest retail electric company has sought to limit the amount of new net-metered solar capacity in its network while pro-solar advocates have set their sights on a Million Roofs campaign that would raise solar electricity from distributed and utility-scale projects to nearly one-fifth of the state's total power use.[22] Free-marketeering Tea Party activists in Georgia have created a Green Tea Coalition, echoing TUSK's support for solar net metering as a matter of unhampered individual choice.

And a high-profile legislative debate in California led to the adoption of a law that extends net-metered solar power to new as well as existing customers through July 2017, but with a cap limiting it to 5 percent of each utility's aggregated peak customer demand.[23]

—∿—

Along with bringing to the surface disparate views on the costs and benefits of distributed solar, the net-metering debate raises much more fundamental questions about the future of electricity production in America—a sector that NRG Energy's CEO David Crane has bluntly called "an industry of Neanderthals." Crane's company, with two-and-a-half million customers, is the nation's largest independent power producer. Its assets run the gamut from coal, gas, and nuclear plants to rooftop solar installations like the ones I toured at Arizona State University and Gillette Stadium. Crane's often brash observations have shaken the electric power industry from within. Though he clearly revels in being the utility sector's bad boy, his words feel genuine when he warns that our nation's big electric power providers must adapt to a much more decentralized, customer-driven model if they are to survive beyond the next few decades. Speaking to a crowded audience at the Massachusetts Institute of Technology, Crane said: "Hopefully before the end of my life, we'll actually be pulling down high-voltage transmission lines around the country." He foresees a new electric order in which electricity consumers will exercise much greater individual choice in the way they purchase and produce the power they need.

In Crane's imagined future, rooftop solar and other distributed power sources will be backed up by compact gas-fired power plants that transform neighborhoods, corporate campuses, and even individual homes into self-sufficient energy islands. When this happens, he asserts, "big centralized power stations will be a thing of the past."[24] Revealing just how far down the power-production food chain his company is willing to go, Crane can point to NRG Home Solar's scramble for a share of US residential PV installations.

Though his company lags behind SolarCity and several competitors in this market niche, Crane is committed to expanding NRG's home-focused presence in California and other solar-friendly states.[25] He can also describe NRG's recent acquisition of a start-up company called Goal Zero, which manufactures handbag-sized, solar-charged battery packs and solar-powered mobile generators for household emergencies.[26]

When I began getting really interested in renewable energy in the 1970s, people who advocated for local energy self-reliance were seen as anti-corporate radicals, communal utopians, survivalists, or all of the above. Breaking free of distant, profit-hungry utility giants ran hand-in-hand with the aspiration to tap seemingly limitless sources of clean energy such as the wind and sun. Today energy autonomy for localities, businesses, public institutions, and even individual households is moving toward the mainstream, espoused by corporate executives like David Crane as well as political conservatives like Barry Goldwater Jr. This reflects a momentous shift away from a status quo that keeps electricity consumers wholly dependent on industrial-scale power plants and a vast network of transmission and distribution lines. Along with the emergence of technologies that can deliver electricity on a much more localized scale, we are developing a whole new vocabulary to represent this shift toward a decentralized energy future. Most benignly, we speak of "distributed energy systems" that can ease, if not eliminate, our dependence on big power plants and the fossil and nuclear fuels they rely on. To reflect the transformative nature of innovations like rooftop photovoltaics, decentralized power storage, and smart energy metering, we refer with excitement or anxiety (depending on our particular perspective) to "disruptive technologies" that will change the way we produce, store, and use electric power. And in a growing number of utility boardrooms, company executives and their advisers warn of a "utility death spiral" in which central generating stations and costly transmission investments lose their value as more and more customers develop their own independent power resources.

The Edison Electric Institute (EEI) sees itself as the voice for

America's investor-owned electric utilities, which serve 220 million Americans in all 50 states and employ more than half a million people. A study recently commissioned by EEI, *Disruptive Challenges: Financial Implications and Strategic Responses to a Changing Retail Electric Business*, takes little comfort in the fact that distributed energy resources make up less than 1 percent of America's total electric load today. The study anticipates a loss in utility revenues in the short term, as customers generate and store more of their power on-site. Investor profits will then decline, making it harder to raise capital for new facilities or necessary upgrades to existing ones. Rate increases may help fill revenue gaps in the short term, but over the longer haul, the higher cost of power will raise the risk of customers' "fully exiting from the grid," causing "irreparable damages to revenues and growth prospects."[27] This is the utility death spiral as EEI foresees it.

Those who ponder the fate of electric utilities in the face of rapid technology changes and expanding consumer choice tell various cautionary tales—of the airlines that evaporated as a result of that industry's deregulation in the 1970s; of Kodak's and Polaroid's failure to make the leap to digital photography; of the phone companies that didn't keep pace with the shift toward mobile telephony and bundled media services. To be sure, EEI is right: distributed energy resources *do* pose a threat to utilities as they are currently structured. So long as utilities view their primary function as pumping electrons through wires rather than satisfying a more diversified set of energy needs, their business model will be vulnerable to the innovations in energy supply, storage, and management that are now penetrating the retail electricity marketplace. That threat may be marginal today, but if a quarter or more of our power needs will someday be met through distributed energy systems, the impact on utility bottom lines will be momentous—*unless* those companies radically reshape their business models.

What exactly might those new business models and services be? The EEI study expresses skepticism about utility programs that offer incentives for energy efficiency and demand-side management; past efforts to recover those investments from ratepayers have not been a pleasant experience, it says. (Of course, that is exactly where some of the biggest progress can be made in cutting our greenhouse gas emissions, and utilities *need* to be working actively with their customers to help them reap those opportunities.) A more promising opportunity for future profits, the study maintains, lies in utilities claiming "ownership of distributed resources with the receipt of an ongoing service fee."[28] Applying this general model, utilities can step into the role now played by SolarCity and other private companies that retain third-party ownership of rooftop solar arrays on homes and businesses, either leasing them to their customers or selling them the electricity produced via power purchase agreements. German utilities, facing much higher levels of solar penetration than most US utilities, are actively considering this option as a way to recover some of the revenues they have lost to distributed solar generation.[29] And in Arizona, two of the state's biggest utilities—APS and Tucson Electric Power—have provisionally won the right to lease rooftop solar arrays to their residential customers.[30]

Another way for utilities to secure a share of the distributed solar realm is to own and operate community solar installations. This avenue is already being pursued by a number of utilities that have built their own community-scale PV installations and have invited customers to buy a share of the power produced by them. As project owners, the utilities—so long as they are privately owned—can take advantage of the federal investment tax credit and other tax benefits attributable to solar energy. They also can count the output from these facilities toward their renewable energy quotas in the states that have adopted renewable electricity standards.[31]

We may not achieve full autonomy from the grid in the foreseeable future, and that probably isn't desirable as a long-term goal in any case. Even if we make a sweeping shift to renewable energy, there will still be notable advantages to maintaining and even expanding the US grid. A highly integrated grid will give us access to prodigious amounts of renewable energy in remote sections of the country where nearby populations are small relative to the available clean energy riches. The sunlight that saturates so much of the Southwest is one such asset; the winds that blow through the Rockies and Great Plains are another. As we draw more fully on these resources, there are major reliability gains to be achieved by balancing the variable output of solar plants and wind farms, taking advantage of natural differences in weather conditions across broad geographical expanses. When the wind blows strong in South Dakota, Oklahoma may be suffering summer doldrums. When the sun is beaming down on the Solana solar thermal power plant in Gila Bend, there may be cloud cover at the California Valley Solar Ranch in San Luis Obispo County. And in the evening, after the sun has set on solar power production, wind farms remain a potent energy resource.

The imperative that we maintain a healthy, resilient grid should not, however, allow us to be complacent about fostering a much higher level of local and sublocal energy self-reliance. Solar photovoltaics are just one of many change agents that call for a fundamental restructuring of the way we produce, store, and consume electricity. We can thank iconoclasts like David Crane for challenging the relevance of centrally produced electricity to the environmental imperatives and technological possibilities of the twenty-first century. We can also thank activists like Barry Goldwater Jr. for championing the role of individual consumers in shaping our electricity future. Just as we need to jettison old assumptions about the role of conventional power plants in meeting our electricity requirements, we need to engage creative minds and principled passions across America's political spectrum as we forge a new set of power dynamics for the nation.

Our Solar Future

"CLIMATE CHANGE IS ONE of the gravest crises our planet has ever faced." With these words, Bristol County district attorney Sam Sutter explained why he had decided to drop criminal charges against two environmental activists who had prevented a freighter from offloading 40,000 tons of Appalachian coal at the Brayton Point Power Station in Somerset, Massachusetts.

In May 2013 Jay O'Hara and Ken Ward anchored a small lobster boat in the freighter's way, an act of civil disobedience intended to highlight the potentially calamitous impacts of coal-burning on our global climate. Predictably, it took the Coast Guard only six hours to lift the anchor and remove the boat, but that was enough time for the two protesters to make a media splash. Facing criminal charges that could have earned them several months' jail time, O'Hara and Ward prepared for a trial in which climate change guru Bill McKibben and former NASA scientist Jim Hansen were to testify about the "necessity" of their action in light of our imperiled global climate. Just before the trial was to begin, District Attorney Sutter made his move. Instead of facing possible prison time, the two protesters were to pay $2,000 apiece in civil fines, covering the costs of removing their vessel from the freighter's path.

What was stunning about Sam Sutter's action was his paradoxical reasoning: he decided to downgrade the criminal charges against O'Hara and Ward precisely because he feared that the judge might actually put them behind bars—a move that, in Sutter's view, would have punished them unduly for their actions. As prosecutor, he couldn't endorse the intentional disruption of ship traffic, but as a concerned citizen, he revered their courage and identified with their cause. "In my humble opinion," he announced to a crowd of reporters and climate activists outside the local courthouse, "the political leadership on this issue has been sorely lacking."[1]

For Sutter and the lobster boat protesters, the fight against climate change is a matter of deep personal conviction. Luckily, it is also a challenge that invites practical engagement on many levels. The search for new energy pathways has led me to far corners of America, first to explore the emergence of wind power as a low-carbon energy resource and more recently to scope out solar power's prospects as a substitute for conventional fuels. I have come away from those explorations with a much deeper belief in the transformation that our nation can achieve if we set the right policies, make smart investment choices, and rally our collective creativity and determination.

With a rooftop solar array now supplying most of our home's power needs and fueling our electric car, I feel cautiously pleased with the steps my family has taken to reduce our carbon footprint. What keeps me from feeling too smug is the knowledge that we as a nation have barely moved the needle in our efforts to dial back America's consumption and combustion of fossil fuels. Hundreds of thousands of American households and businesses may have made the leap to solar energy in recent years, but tens of millions more need to make the same move if rooftop photovoltaics are to provide a fifth or more of our nation's electricity—a level of solar integration that the National Renewable Energy Laboratory sees as feasible using currently available technology.[2] I have also marveled at the utility-scale solar installations that have begun to spread across farmlands and the open desert, and have witnessed the repurposing of polluted industrial brownfields. Yet these installations represent just a fraction of the vast solar potential that remains untapped in our open spaces and urban areas.

To move from inspiring but modest innovation to wholesale transformation, we need to map out a pathway to the future. The states that have set renewable electricity standards have made a start, but even the most far-reaching of those standards—California's—will only get us to 33 percent reliance on renewable resources for our electric power needs by 2020. That target is a good one for California in the near term, but where do we as a nation need to be in the longer term, and what are the other key elements in a climate action plan that would be worthy of America's stature as a leading global industrial power?

If predominant scientific assessments are at all accurate, we will have to shift four-fifths of our global electricity production away from fossil fuels by mid-century to ward off some of the worst effects of global warming: widespread population displacement, ruinous droughts, food shortages, and species extinction.[3] To be sure, America is only one player—though a critically important one—in a world where rapid economic development and runaway population growth are driving up overall energy use. These trends are happening precisely at a time when we need to be trimming global energy consumption and shifting toward more sustainable energy feedstocks.

The tepid results of successive UN climate treaty negotiations suggest that the United States and other industrialized nations would be foolhardy to wait for binding global commitments before developing and implementing their own climate action plans. To shape an effective climate strategy, American politicians and policy makers will need to set aside their facile rejection of long-range national planning as a tool reserved for fascist dictatorships and communist regimes. We need to take heart in the enormous strides made by other Western democracies that have been far less wary of venturing down the path toward coherent national energy planning. We need to embrace and then vigorously pursue a set of measures that will dial down the energy waste that plagues virtually every sector of our economy and dial up our use of resources that will allow us to move away from carbon-based fuels.

Fortunately, there are a few great models of broad-based energy planning that can inspire a more visionary American energy agenda. One is the European Union, which has already shown that it's possible to achieve vigorous economic growth while significantly lowering greenhouse gas emissions. By 2020, the EU has committed to cutting the overall greenhouse gas emissions of its 28 member nations by 20 percent, using 1990 emissions as a baseline. Individual EU states face specific targets based on a number of economic and demographic factors, but each nation is free to develop and implement its own means of achieving its mandatory target. The results, so far, have been impressive. By 2012 the EU had trimmed back its greenhouse gas emissions by 19.2 percent, at the same time that it achieved a 45 percent overall growth in gross domestic product. The European Commission anticipates that by 2020, EU greenhouse gas emissions will have dropped to 22.2 percent below 1990 levels.[4]

Over the longer term, the EU has adopted greenhouse gas reduction targets that are much more ambitious, culminating in 80 to 95 percent lower emissions by 2050. As with any long journey, interim milestones and travel tips are very helpful. To support member nations as they chart out their own routes to this goal, the European Commission has prepared a 2050 Roadmap that identifies steps that can be taken across multiple sectors. It has also set interim milestones for shrinking the EU's carbon footprint: a 40 percent reduction in greenhouse gases by 2030 and a 60 percent drop in those emissions by 2040.[5]

Of all EU nations, Denmark is making the biggest strides toward these milestones. By 2035 all Danish electricity and heating needs are to be met with renewable resources, and all of the country's energy supply, including fuel for transportation, is to be based on renewable energy by 2050.[6] This may sound far-fetched to carbon-reliant Americans, but it hardly daunts a nation that already gets nearly 40 percent of its power from wind energy[7] and expects to achieve a 40 percent reduction in economy-wide greenhouse gas emissions by 2020.[8]

If the United States were to adopt a national climate action plan, we would be wise to emulate the EU in setting both an aggressive long-range target and binding interim milestones. Looking at all sectors of the economy, we would take bold steps to improve the energy performance of our buildings, appliances, industrial processes, and agricultural practices. We would ramp up automobile fuel efficiency and invest in lower-impact means of moving people and goods across town and around the country. We would redesign our cities to make them less energy-wasteful and car-dependent. And we would orchestrate a major shift away from carbon-based fuels in the production of electricity.

Solar energy will surely have a place in this transformed energy constellation, but just how significant might its role be? We know that technology is no longer the barrier; mass-produced photovoltaic panels are now a standard commodity in the global marketplace, and their quality and durability are improving constantly. Concentrating solar power is gaining ground as well, even if its progress has been slowed by higher costs and concerns about wildlife impacts. We also know that the sun itself is both superabundant and highly adaptable to diverse settings, from household and commercial rooftops to industrial brownfields and the open desert.

One central task will be setting the policy signals that will economically sustain solar technology during the coming years. The 30 percent investment tax credit has been critical to creating a level playing field for solar power in an industry long dominated by carbon-based and nuclear fuels. Renewable electricity standards and the various other state and local incentives discussed in chapter 10 have also built solar power's early momentum. If we maintain the right policy signals, we stand a very good chance of fulfilling the National Renewable Energy Laboratory's forecast that half the nation's electricity could come from solar and wind energy by 2050. Germany, with its gray weather, draws 7 percent of its power from the sun today, less than two decades after launching its 100,000 Roofs Program in 1998, when photovoltaics were still very costly.[9] Today market conditions have greatly improved, and the technology itself is steadily advancing. Given America's superabundant land resources and much more favorable weather conditions, we should have little difficulty tripling or even quadrupling Germany's level of reliance on the sun during the next two decades *if* we were to make this a national priority.

The sun has a central role to play in setting America on course toward a low-carbon energy future, but it is no stand-alone panacea. A robust American energy portfolio will draw upon diverse energy resources as well as much more sophisticated management of the energy we consume. Utilities may grumble about the erosion of their guaranteed customer base; coal and gas interests may balk at new technologies cutting into their share of the US power market. Thankfully, these forces of recalcitrance and resistance are not the only voices being heard today. A new generation of entrepreneurs and engineers is weighing in with vision and creativity, opening up new frontiers for solar development, and politicians across the political spectrum are waking up to the opportunities that accompany a genuine commitment to renewable energy. The tools for advancing a more sustainable energy future are within reach. It is our obligation and privilege to use them.

Acknowledgments

JUST AS HARNESS THE SUN begins in our home, so will my thanks. My wife, Tamar, has been my most ardent and steadfast supporter through all stages of researching and writing this book. An enthusiastic collaborator in shrinking our family's carbon footprint, she has also been my reader of first resort, offering just the right balance of encouragement and editorial guidance as the book has taken shape. My daughter Tali spent her first weeks as a college graduate reading these pages, pinpointing the spots where key issues demanded sharper focus and filling some important data gaps. My younger daughter, Maya, still in college, has opened my mind to the social media world, convincing me of its value as a gateway to future environmental leaders.

I am grateful to my fabulous agent, Colleen Mohyde at the Doe Coover Agency, for her enthusiasm about this book and for landing it at Beacon Press, where progressive ideas are nurtured with refreshing determination. At Beacon, I feel hugely fortunate to have worked once again with Alexis Rizzuto, editor of my first renewable energy exploration, *Harvest the Wind*. Along with her wonderfully keen editorial eye, Alexis has an activist's deep commitment to tackling the climate change conundrum. To the rest of the Beacon team, including Marcy Barnes, Travis Dagenais, Tom Hallock, Alyssa Hassan, Susan Lumenello, Pam MacColl, and Will Myers, I owe my deep thanks for carrying this book so masterfully from manuscript to publication.

Rosemary Ahern, another cherished veteran of *Harvest the Wind*, has been a smart and sensitive guide to the crafting of this book, making sure that my writing keeps general readers tuned in. My sister Sally Bliumis Dunn, a gifted poet and teacher, inspired me to bring real people to the fore in the telling of this story. Jim Rogers at the Union of Concerned Scientists and Rob Sargent at Environment America combed the manuscript for accuracy and provided valuable perspective on solar power's transformative potential. Dwayne Breger and Howard Bernstein of the Massachusetts Department of Energy Resources carefully reviewed sections on state policy. Christian Honeker of the Fraunhofer Center for Sustainable Energy Systems and Ben Santarris of SolarWorld honed my exploration of solar panel manufacturing, testing, and life-cycle management.

I interviewed more than 120 people in the course of my research for *Harness the Sun.* I want to thank all of them for sharing their knowledge and insights with me. Among those who do not appear in the text but who opened many doors and provided a wealth of information and ideas, I owe special thanks to Elizabeth Ackerman and B. Scott Hunter of the New Jersey Board of Public Utilities; Joseph Cabral and Jason Caudle with the City of Lancaster (CA); David Colt of Global Power Finance; Brianna Gianti, Tom Kimbis, Hannah Panek, and Shawn Rumery at the Solar Energy Industries Association (SEIA); Tanuj Deora at the Solar Electric Power Association (SEPA); Leonard Gold of the Gila River Indian Community Utility Authority; Amy Heuslein and her colleagues at the US Bureau of Land Management; Enesta Jones in the US Environmental Protection Agency's Office of Media Relations; Trina Lindsey at the Bedminster (NJ) Township Office; Kern County (CA) planning director Lorelei Oviatt; Andi Plocek and Mary Grikas at SolarReserve; Constellation Energy's Brendon Quinlivan; Mark Rafferty and Dave Young at the Facility Management Group; Amit Ronen at George Washington University's Solar Institute; Tim Roughan and Justin Woodard at National Grid; DC SUN's Anya Schoolman; PSE&G's Francis Sullivan; and Karin Wadsack at Northern Arizona University.

Other unsung heroes are the members of my dedicated crew of research assistants: Sabina Ambani, Colin Cushman, Caitlin Furio, Miriam Haviland, Arielle Herzberg, Brandon Jewart, Janée Johnson, Anisha Kalyani, Molly Meyer, and Margaret Therrien. Thanks, too, to Samara Gordon, for launching me on Twitter.

I have also been blessed with dear friends and family members whose interest and encouragement have motivated me every day. To Joseph Blasi, David and Lakshmi Bloom, Mel Brown, Jonny and Lauren Garlick, Mike and Cathy Gildesgame, Harold Grinspoon, Evan Kaizer, Robin Kramer, Art Kreiger and Rebecca Benson, Michael Shiner, my book-writing mentor Larry Tye, and my brother Jim Warburg, thank you. I hope that this book, in some small way, helps secure a safer, more sustainable world for all of our children and grandchildren.

Notes

INTRODUCTION

1. Warburg, *Harvest the Wind*.

2. Shawn Rumery, research manager, Solar Energy Industries Association (SEIA), e-mail to author, March 16, 2015, citing SEIA/GTM Research, *U.S. Solar Market Insight Report—2014 Year in Review*, http://www.greentechmedia.com/research/ussmi.

3. Solar Foundation, *National Solar Jobs Census 2014*, p. 1, http://thesolarfoundation.org/.

4. In addition to electricity generation, common solar energy applications include water heating for household, commercial, and industrial use; lighting; space heating and cooling through active and passive measures; disinfection; and water purification.

5. SEIA, Solar Data Cheat Sheet, accessed March 15, 2015, http://www.seia.org.

6. SEIA/GTM Research, *2014 Year in Review*, Executive Summary, fig. 1.2, http://www.seia.org/.

7. Solar power (photovoltaic and thermal) produced 18,321 gigawatt-hours of electricity in 2014, or 0.45 percent of US power generation totaling 4,092,935 gigawatt-hours. This was more than double its contribution to our power supply in 2013, when 9,036 gigawatt-hours of electricity came from solar. Energy Information Administration, *Electric Power Monthly*, February 2015, http://www.eia.gov.

8. Lopez et al., *U.S. Renewable Energy Technical Potentials: A GIS-Based Analysis*, table 12. This table estimates that 800 terawatt-hours (800 trillion watt-hours) of electricity per year could come from rooftop photovoltaics, out of a total solar technical potential of 399,700 terawatt-hours. US retail electricity sales in 2010 added up to 3,754 terawatt-hours.

9. Hand et al., eds., *Renewable Electricity Futures Study, Vol. 1: Exploration of High-Penetration Renewable Electricity Futures*, xvii.

10. SEIA, Solar Data Cheat Sheet, updated March 6, 2015, http://www.seia.org/. Cumulative installed US solar capacity reached 20.5 gigawatts by the end of 2014, according to SEIA/GTM Research, *2014 Year in Review*.

11. At the end of 2013, cumulative installed PV capacity was 12.1 gigawatts and cumulative CSP capacity was 918 megawatts, adding up to just over 13 gigawatts. In 2014, 6.201 gigawatts of new PV capacity and 767 megawatts of concentrating solar power (CSP) were installed, bringing the year's total to 6.968 gigawatts. SEIA/GTM Research, *2013 Year in Review; 2014 Year in Review*.

12. Barbara Rose Johnston et al., "Uranium Mining and Milling," in Smith and Frehner, eds., *Indians & Energy: Exploitation and Opportunity in the American Southwest*, 128.

13. IPCC, *Climate Change 2014: Mitigation of Climate Change, Summary for Policymakers*, 21, http://report.mitigation2014.org/spm/ipcc_wg3_ar5_summary-for-policymakers_ap

proved.pdf. The IPCC leaves open the possibility that fossil fuels might be used if adequate means of capturing and sequestering carbon are developed and implemented. Nuclear power is also included within the IPCC's definition of low-carbon electricity. The IPCC's prescribed fuel transitions correspond to the achievement of "low-stabilization levels" for greenhouse gas emissions of 430–530ppm CO_2 equivalent.

CHAPTER ONE: OUR HOUSE, YOUR HOUSE

1. See chapter 8 for a discussion of the impact of temperature, among other factors, on PV performance.

2. Under the Massachusetts RPS, "new" renewable electricity facilities are defined as those that began operating after January 1, 1998. This avoids letting older plants, such as hydro dams, satisfy the requirements. As part of the overall RPS obligation, a solar "carve-out" provision requires at least 1.6 gigawatts of solar generating capacity to be installed by 2020. A description of the Massachusetts RPS can be found at http://www.dsireusa.org/. Detailed information on the state's solar requirements and SREC program is available at http://www.mass.gov/eea/energy-utilities-clean-tech/renewable-energy/solar/rps-solar -carve-out/about-the-rps-solar-carve-out-program.html.

3. Solar installers generally warn customers that inverters, which convert DC current generated by the solar cells to AC current compatible with electric distribution, may need to be replaced during the contract period. Warranties on inverters generally expire after five to ten years.

4. Hoen et al., *Selling into the Sun*, 29–30. In this study, a 3.6 kilowatt PV array was defined as average-sized. An earlier study looking at homes in San Diego County found an average price premium of 3.5 percent for PV-equipped homes. See Dastrup et al., *Understanding the Solar Home Price Premium*, 5–8.

5. US Department of Energy (SunShot), "Sharing the Sun: Solar Power for Tenants and Solar Disadvantaged Homeowners," 2013, http://catalyst.energy.gov/.

6. The legal status of third-party-owned solar installations remains ill-defined in many state jurisdictions. Only a handful of states (e.g., California, Colorado, Nevada, New Jersey, and Oregon) have laws that explicitly authorize third-party ownership of solar installations. In several other states, legal challenges have sought to require that companies using third-party PPAs be defined as sellers of electricity, subject to the often onerous regulations governing public utilities. Leases, involving the rental of equipment rather than the sale of electricity, have raised fewer objections. See Kollins et al., *Solar PV Project Financing*.

7. Cumulative residential rooftop installations by the end of that year totaled more than 598,000. By that time, commercial PV installations accounted for another 46,000 systems, and there were roughly 1,000 operating utility-scale systems. Shawn Rumery, research manager, Solar Energy Industries Association (SEIA), e-mail to author, March 16, 2015, citing SEIA/GTM Research, *2014 Year in Review*.

8. "Sunrun and PV Solar Report Announce Third-Party-Owned Solar Generated More Than $900 Million for California in 2012," Business Wire, February 13, 2013, http://www.businesswire.com/. In California, third-party-owned PV arrays accounted for 69.1 percent of new installations in Q1 2014. SEIA/GTM Research, *U.S. Solar Market Insight Report—Q1 2014*, fig. 2.11.

9. SEIA/GTM Research, *U.S. Solar Market Insight Report—Q1 2014*, fig. 2.11. In Q4

2012, third-party-owned installations in Arizona hit 90 percent. In Colorado, they reached 91.2 percent in Q1 2013. In New Jersey, 93.1 percent of new home solar installations were third-party-owned in Q4 2013.

10. The original legislation establishing a mandate for solar on low-income housing was California Assembly Bill 2723 (2006). Funding for the program was expanded by Assembly Bill 217 (2013). Detailed reportage on the SASH program is available online at the California Public Utilities Commission website, http://www.cpuc.ca.gov/.

11. Cathleen Monahan, SASH program officer, GRID Alternatives; Mara Meaney-Ervin, Bay Area development officer, GRID Alternatives; Ron Griffin, solar installation supervisor, GRID Alternatives; and Liying Huang, homeowner, Bayview district, San Francisco: interviews with author, April 23, 2013.

12. Brent Haverkamp, CEO, Haverkamp Properties, Ames, Iowa, phone interview with author, September 11, 2014.

13. Tom Sweeney, chief operating officer, Clean Energy Collective, panel participant, "Exploring Community Solar Programs," Solar Power International, Chicago, October 24, 2013.

14. This figure is extrapolated from SEIA's estimate of the average number of California homes powered by a megawatt of solar PV: 216. SEIA estimates that the national average number of homes powered by a megawatt of solar PV is 164. See "What's in a Megawatt?," http://www.seia.org/, accessed January 6, 2015. The US Environmental Protection Agency assumes a lower national average of 146 homes powered by a megawatt of solar PV. Enesta Jones, Office of Media Relations, EPA, e-mail to author, April 29, 2014.

15. California Senate Bill 43 (2013), http://leginfo.legislature.ca.gov/. See also Interstate Renewable Energy Council, *Model Rules for Shared Renewable Energy Programs*, 2013, http://www.irecusa.org/.

16. Between June 2, 2013, and October 22, 2014, the car was driven 10,173 miles. During that period, the tank was refilled six times, for a total of 66.6 gallons. According to the C-Max's onboard computer, the car operated on electricity for 80 percent of the traveled miles.

17. This online monitoring program, SolrenView, supplies real-time data on the output of our solar arrays and provides useful comparative metrics such as the ones cited here. A second web-based monitoring program, run on Enphase Enlighten software, shows me the exact, real-time output of every individual panel, making it easier to pinpoint specific sources of system underperformance.

18. At their peak, SRECs sold for as much as $540. By the end of 2013, they were selling for less than half that amount. To avoid an even more severe drop in SREC payments, Massachusetts has an auction mechanism that sets a "soft" floor price for SRECs in years of oversupply.

19. In November 2014 our electricity provider NStar (now Eversource) petitioned for a 29 percent rate increase starting in January 2015; two months earlier its competitor National Grid had successfully filed for a 37 percent rate increase. Jack Newsham, "NStar Seeks 29 Percent Hike in Electric Rates," *Boston Globe*, November 7, 2014.

20. Paul Israel, president, Sunlight Solar Energy, phone interview with author, November 21, 2013.

CHAPTER TWO: BALLFIELDS AND BOXTOPS

1. In addition to the solar installation at Patriot Place, NRG has built PV arrays at the Redskins' FedEx Field in Landover, MD; the Eagles' Lincoln Financial Field in Philadelphia; the Giants' MetLife Stadium in the Meadowlands; and the 49ers' Levi's Stadium in San Francisco. A sixth array at NRG Stadium, home of the Houston Texans, is expected to enter operation by the end of 2015.

2. For a listing of NRG's generation assets, see http://www.nrgenergy.com/pdf /Projectlist.pdf.

3. Thomas D. Gros, chief customer officer, NRG, and president, NRG Solutions, interview with author, October 27, 2013.

4. "A Big Green Apple Campus," *Energy Prospects West*, October 29, 2013.

5. Google's renewable energy investments are documented at http://www.google.com /green/energy/.

6. Solar Energy Industries Association (SEIA), *Solar Means Business 2014*, 8.

7. John Finnigan, "How Utilities Can Adapt When Big-Box Retailers Go Solar," *GreenBiz.com*, August 16, 2013, http://www.greenbiz.com/.

8. See Institute for Local Self-Reliance, *How the Walton Family Is Threatening Our Clean Energy Future*.

9. Richard Bass, president, Cardinal Shoe, interview with author, June 19, 2013.

10. See National Renewable Energy Laboratory (NREL), *Section 1603 Treasury Grant Expiration: Industry Insight on Financing and Market*.

11. Kathleen C. Doyle, CEO and founder, FireFlower Alternative Energy, interview with author, June 20, 2013.

12. Under revised Massachusetts regulations issued in 2014, differential SREC values are assigned to particular types of solar generation, depending on the priority that state policy makers have given them. Household-scale installations and shared community generation units receive full valuation, whereas larger building-mounted systems and ground-mounted systems that generate 67 percent or more of their power for on-site use receive a 0.9 valuation. Lower valuations are assigned to systems built on landfills or brownfields, or ground-mounted systems that supply less than 67 percent of their power for on-site use. See 225 CMR 14.00 (April 24, 2014), summarized online at http://dsireusa.org/.

13. John Perlin, *Let It Shine*, 310–16.

14. See NREL, Dynamic Maps, GIS Data & Analysis Tools, http://www.nrel.gov /gis/solar.html. Maps on this website show solar resource information by state and by solar technology (photovoltaics and concentrating solar power).

15. See fig. 4 in chapter 1 for a ranked listing of cumulative solar PV capacity for the top thirty states.

16. NJ Senate Bill 1925 (2012) is summarized at http://www.dsireusa.org/incentives /incentive.cfm?Incentive_Code=NJ05R.

17. Avi Avidan, managing member, and Josh Avidan, member, Avidan Management LLC, Edison, NJ, interview with author, June 24, 2013.

18. In the fourth quarter of 2014, PV modules sold in the United States averaged $0.73 per watt. SEIA/GTM Research, *2014 Year in Review*, Executive Summary, fig. 2.8, http:// www.seia.org/.

19. Alan Epstein, president and COO, KDC Solar, interview with author, June 26, 2013. See individual project descriptions at http://www.kdcsolar.com/.

20. New Jersey assemblyman Upendra Chivukula, interview with author, June 25, 2013. On September 22, 2014, Chivukula resigned his Assembly seat to become a commissioner on the New Jersey Board of Public Utilities.

21. Lyle Rawlings, CEO, Advanced Solar Products, and vice president, Mid-Atlantic Solar Energy Industries Association, phone interview with author, August 21, 2013. For more information on NJ FREE, see http://mseia.net/solar/nj-free/.

22. See ASU Solar at https://cfo.asu.edu/solar/.

23. Jean Humphries, director of design and support services, capital programs management group, Arizona State University, interview with author, December 4, 2013.

24. As of October 10, 2014, 684 colleges and universities had signed on to this commitment. For an updated list and information on campus plans, see http://www.presidents climatecommitment.org/.

25. A. Karl Edelhoff, senior project manager, Capital Programs Management Group, Arizona State University, interview with author, December 4, 2013.

26. These rankings are current as of March 25, 2015. See updated listing at http://www .aashe.org/resources/campus-solar-photovoltaic-installations/top10/.

27. The Rutgers Solar-to-Vehicle (S2V) Project is described at http://www.rci.rutgers .edu/~dbirnie/Solar2Vehicle/.

28. Dunbar Birnie III, professor of ceramic engineering, Rutgers University, interview with author, June 27, 2013.

29. Green Sports Alliance, "Rutgers Powers Campus with Largest Solar Canopy System in Nation," September 13, 2013, http://greensportsalliance.org/.

30. Amit Ronen, director, GW Solar Institute, interview with author, April 15, 2014.

31. GW Office of Media Relations, "Fact Sheet—Capital Partners Solar Project," http://sustainability.gwu.edu/. This project was announced on June 24, 2014.

32. For further details about GW's Climate Action Plan, see http://sustainability.gwu .edu/.

33. White House, Office of the Press Secretary, "Fact Sheet: Reducing Greenhouse Gas Emissions in the Federal Government and Across the Supply Chain," March 19, 2015, http://www.whitehouse.gov/.

34. ACORE, *US Department of Defense & Renewable Energy*, 2.

35. Under the John Warner National Defense Authorization Act for Fiscal Year 2007, P.L. 109-364 (2006), Sec. 2852, a goal was set for DOD to produce or procure not less than 25 percent of its total electricity needs from renewable energy sources during FY 2025 and thereafter.

36. See ACORE, *US Department of Defense & Renewable Energy*.

37. The lower estimate appears in US Department of Defense, *Annual Energy Management Report, Fiscal Year 2012*, 6. The higher estimate appears in the fact sheet announcing the president's executive order on March 19, 2015.

38. North American Development Bank, "Davis-Monthan Air Force Base Celebrates Completion of a 16-MW Solar Plant," press release, February 13, 2014, http://www.nadb .org/.

39. "13.78-Megawatt SunPower Solar Power Plant at NAWS China Lake Begins Oper-

ations, Expected to Reduce Costs by $13 Million," PR Newswire, October 19, 2012, http://prnewswire.com/.

40. Georgia Power, "Georgia Power to Bring 90 MW of Solar to Army Bases," press release, May 15, 2014, http://www.georgiapower.com/.

41. DOE, Office of Electricity Delivery & Energy Reliability, *Hurricane Sandy-Nor'easter Situation Report #13*, December 3, 2012, http://www.oe.netl.doe.gov/.

42. Chelsea J. Carter, "Arkansas Man Charged in Connection with Power Grid Sabotage," CNN, October 12, 2013, http://www.cnn.com/; "Vandalism at San Jose PG&E Substation Called 'Sabotage,'" CBS News–SF Bay Area, April 16, 2013, http://sanfrancisco.cbslocal.com/.

43. Rebecca Smith, "Assault on California Power Station Raises Alarm on Potential for Terrorism," *Wall Street Journal*, February 5, 2014.

44. Lisa Ferdinando, "Fort Bliss Unveils Army's First Microgrid," Army News Service, May 17, 2013, http://www.army.mil/.

CHAPTER THREE: LOCAL COMMUNITIES CAPTURE THE SUN

1. R. Rex Parris, mayor, Lancaster, CA, interview with author, May 1, 2013. Unless otherwise indicated, all other references to Mayor Parris are from this interview.

2. Lancaster's installed solar capacity is reported in GoSolar California, "Geographical Statistics," updated October 15, 2014, http://www.californiasolarstatistics.ca.gov/. To equate this to the power consumption of average California households, I rely upon data supplied by the Solar Energy Industries Association (SEIA), "What's in a Megawatt?" http://www.seia.org/ (estimating that the power needs of 216 average California households can be met by a megawatt of solar PV installed in the state).

3. Heather Swan, project coordinator, Lancaster Power Authority, and Jason Caudle, deputy city manager, Lancaster, CA, interview with author, May 1, 2013.

4. City of Lancaster Solar Statistics, updated February 28, 2014 (provided by Jocelyn Swain, associate planner, City of Lancaster).

5. See fig. 1 in "Note on Terminology." Since the average number of California homes powered by a megawatt of solar photovoltaics is 216, the average California household requires about 4.6 kilowatts of installed solar capacity to supply 100 percent of its power.

6. Lancaster, California Code of Ordinances, Title 15.28—Energy Code, Sec. 110.11: Mandatory Requirements for the Implementation of Solar Energy Systems (2013), http://library.municode.com/.

7. The Federal Trade Commission has raised a related concern about companies that generate renewable electricity and then sell renewable energy certificates for all of that electricity. In such cases, the FTC cautions: "It would be deceptive for the marketer to represent, directly or by implication, that it uses renewable energy." *Code of Federal Regulations*, Guides for the Use of Environmental Marketing Claims, title 16, sec. 260.15(d), rev. October 11, 2012, http://www.ftc.gov/. The FTC's guidance does not appear to be targeted at noncommercial entities, but its warning about double-counting is pertinent here.

8. "Munis Score Solar PPA Savings in Renegotiations with Silverado Power," *California Energy Markets*, March 21, 2014. These two facilities are the Antelope Big Sky Solar Project and Summer Solar Project, initially developed by Silverado Power, which merged with sPower in February 2014.

9. Cal Rev. & Tax Code, sec. 73 (2012), http://www.leginfo.ca.gov/.

10. Jocelyn Swain, associate planner, City of Lancaster, phone interviews with author, March 27, 2014, and November 12, 2014.

11. "Bush Sr. Apologizes for Marin Hot Tub Slam," February 28, 2002, *Los Angeles Times*, http://articles.latimes.com/.

12. California Assembly Bill 117, Ch. 838 (2002), http://www.leginfo.ca.gov/.

13. A concise history of this entity appears in Susan Bryer et al., *Community Choice Aggregation: Lessons Learned from Marin Energy Authority*, Leadership Institute for Ecology and the Economy, April 2011, http://ecoleader.org/.

14. California Public Utilities Commission (CPUC), *Report to the Legislature: Issues and Progress on the Implementation of Community Choice Aggregation*, 3rd quarter report, July 31, 2011, 17. These and other PG&E expenditures devoted to opposing community choice aggregation initiatives are documented in the CPUC's quarterly reports, available online at http://www.cpuc.ca.gov/.

15. Paul Clanon, executive director, CPUC, letter to Brian Cherry, vice president, Regulatory Relations, PG&E, May 12, 2010, http://www.cpuc.ca.gov/NR/rdonlyres/677EC1E6-3420-42A8-BBF0-B8E69859A00E/0/PGELetter051210.pdf.

16. Dawn Weisz, executive officer, Marin Clean Energy (MCE), interview with author, April 24, 2013. Unless otherwise indicated, all other references to Dawn Weisz are from this interview.

17. MCE, Board of Directors Meeting, April 4, 2013, http://marincleanenergy.org/.

18. MCE, *MCE Integrated Resource Plan Annual Update*, November 2013, fig. 3, http://marincleanenergy.org/. If large-scale hydroelectricity is included, renewable energy reaches 60 percent of the electricity in MCE's portfolio.

19. Brett Wilkison, "Sonoma Clean Power Lays Out Green Energy Goal," (Sonoma, CA) *Press Democrat*, October 17, 2013, http://www.pressdemocrat.com/.

20. In addition to the environmental issues surrounding many large hydro dams, California and several other states do not allow large-scale hydro to count toward RPS compliance because of the superabundance of hydropower, particularly in the West. The intent of the RPS is to stimulate the development of new renewable energy resources, not to credit technologies that are already well developed.

21. See CPUC, "Net Surplus Compensation (AB 920)," last modified October 4, 2011, http://www.cpus.ca.gov/.

22. For a description of this initiative, see San Joaquin Valley Power Authority, *Community Choice Aggregation Implementation Plan and Statement of Intent*, Modification No. 3, February 2009, http://cleanenergyus.org/.

23. CPUC, *Report to the Legislature: Issues and Progress on the Implementation of Community Choice Aggregation*, July 31, 2011, p. 17.

24. John Coté, "Lee Decides to Back S.F. Clean Power Program, but with Conditions," *SFGate*, January 26, 2015, http://www.sfgate.com.

25. Leora Broydo Vestel, "Sonoma Clean Power Looks to Future as Expansion Continues," *California Energy Markets*, January 16, 2015.

26. "Enrollment in Sonoma Clean Power Better Than Forecast," *California Energy Markets*, March 28, 2014. See also Sonoma Clean Power website, https://sonomacleanpower.org/.

27. "Palo Alto Electricity—Now 100% Carbon Free!," http://vimeo.com/61055308. For further details on Palo Alto's steps toward carbon neutrality, see *City of Palo Alto City*

Council Staff Report, Electric Supply Portfolio Carbon Neutral Plan, March 4, 2013, http://www.cityofpaloalto.org/.

28. "Palo Alto Approves Major Solar Power Purchase Agreements," *Public Power Weekly*, July 28, 2013, http://www.naylornetwork.com/.

29. San Francisco Public Utilities Commission (SFPUC), *2011 Updated Electricity Resource Plan*, 36, http://sfwater.org/. Electricity supplied by PG&E accounts for about 75 percent of the city's power consumption.

30. Charles Sheehan, communications manager, Power Division, SFPUC, interview with author, April 22, 2013.

31. Diane Cardwell, "A Bet on the Environment," *New York Times*, September 3, 2013.

32. Yuliya Chernova, "Want to Invest a Few Hundred Bucks in a Solar Project? Mosaic Opens Crowdfunding Platform," *Wall Street Journal*, January 8, 2013.

33. Andrew Herndon, "Solar Mosaic's Crowdfunding Beats Treasuries With 4.5% Return," *Bloomberg Business*, January 7, 2013, http://www.bloomberg.com/.

34. Aarti Shahani, "Solar Company Can 'Crowd Fund' $100 Million, Regulators Say," KQED, April 8, 2013, http://www.kqed.org/.

35. Matthew Morrissey, former executive director, NBEDC, interview with author, November 18, 2014. Unless otherwise indicated, all references to Matthew Morrissey are based on this interview.

36. Under the Massachusetts Green Communities Act, the City of New Bedford and the New Bedford Economic Development Council can be treated as separate municipal entities, thereby entitling each to draw up to 10 megawatts of solar power as the "off-takers" or consumers of net-metered electricity.

37. Eric Graber-Lopez, Partner, BlueWave Capital, phone interview with author, November 22, 2013. See also Matt Camara, "Rochester Planning Board Unanimously Approves Little Quittacas Solar Project," (New Bedford, MA) *Standard-Times*, January 9, 2013, http://www.southcoasttoday.com/.

38. John P. DeVillars, managing partner, BlueWave Capital, interview with author, November 5, 2013. All quotes of John DeVillars are from this interview.

39. Massachusetts has a special "carve-out" for solar power that creates a market for Solar Renewable Energy Credits, discussed in chapter 1. For more on this policy, see Massachusetts Executive Office of Energy and Environmental Affairs, "Current Status of the Solar Carve-Out Program," http://www.mass.gov/, accessed January 6, 2015.

40. Max Chafkin, "Entrepreneur of the Year, 2007: Elon Musk," *Inc.*, December 1, 2007, http://www.inc.com/.

41. *BlueWave Capital Overview*, October 2013, provided to author by John P. DeVillars.

42. *BlueWave Overview*; Graber-Lopez interview.

43. Phil Cavallo, president and CEO, Beaumont Solar Co., interview with author, April 7, 2014. All quotes of Phil Cavallo are from this interview.

44. For a survey of leading city-based solar programs, see Environment Massachusetts, *Shining Cities*.

CHAPTER FOUR: WASTELANDS REDEEMED

1. For these and other KDC projects, see http://www.kdcsolar.com/.

2. Alan Epstein, president and chief operating officer, KDC Solar LLC, interview with author, June 26, 2013.

3. Since 1964, New Jersey property owners have needed to show a minimum of $500 in annual income from farming to qualify for a property tax exemption that can run as high as 98 percent of normal rates. In 2013, Governor Christie signed a new law, Senate Bill 589 (P.L. 2013, c.43), http://www.njleg.state.nj.us/, raising this threshold to $1,000 for the first five acres, plus $5 for every additional acre. See Michael Williams, "N.J. 'Fake Farmers' Bill Signed into Law, Updating Decades-Old Assessment Act," *South Jersey Times*, April 21, 2013, http://www.nj.com/.

4. Township of Bedminster, Ordinance No. 2013–013, Sec. 2, http://www.bedminster.us/.

5. Master Plan—Bedminster Township, January 2003, rev. January 2005, 2, 5, http://www.bedminster.us/.

6. Steven Parker, mayor, Bedminster, phone interview with author, July 2, 2013.

7. Bob Marshall, assistant commissioner for sustainability and green energy, NJ Department of Environmental Protection, interview with author, June 26, 2013.

8. After months of delay, the Bedminster land use board's first hearing was held on October 1, 2013. More than a year later, the hearings were ongoing with no resolution in sight. Transcripts can be found on the township's website at http://www.bedminster.us/.

9. New Jersey Senate Bill 1925 (2012), amending Sec. 2(t)(1) of N.J.P.L. 1999, c. 23 (C.48:3–87), http://www.njleg.state.nj.us/.

10. Tom Johnson, "State Approves PSE&G's $500 Million Solar-Energy Initiative," *NJ Spotlight*, May 30, 2013, http://www.njspotlight.com/.

11. Paul Morrison, project manager, Solar4All, PSE&G, interview with author, July 16, 2013.

12. PSE&G Newsroom, "Meadowlands Landfill Solar Farm Dedicated," May 8, 2012, http://www.pseg.com/.

13. Todd Hranicka, solar energy director, PSE&G, interview with author, July 16, 2013.

14. Bjorn B. Jensen, "Brownfields to Green Energy: Redeveloping Contaminated Lands with Large-Scale Renewable Energy Facilities," master's thesis submitted to the Department of Urban Studies and Planning, MIT, June 2010, 90.

15. Deborah Sawyer, founder and CEO, Environmental Design International, interview with author, July 16, 2013.

16. After spending $800,000 on site cleanup prior to Exelon's arrival, the city reportedly allocated $200,000 for further remediation concurrent with the solar facility's development. This, too, was financed by the city, but it was to be reimbursed through Exelon's first two years of rent payments. Jensen, "Brownfields to Green Energy," 90.

17. Carrie Austin, alderman, 34th Ward, City of Chicago, interview with author, July 16, 2013. All quotes of Alderman Austin are from this interview.

18. For more information on this initiative, see ELPC, "Transforming Brownfields into Brightfields," http://elpc.org/b2b/.

19. Kari Lydersen, "Chicago Steelmaking: Dead but Not Forgotten," *Washington Post*, December 27, 2004, http://www.washingtonpost.com/.

20. Howard Learner, president, Environmental Law and Policy Center, phone interview with author, August 7, 2013.

21. The 33 percent renewable energy target was established earlier, under Executive Order S-14–08, but Executive Order S-21–09 set a deadline for the California Air Resources Board to develop the regulations implementing this target.

22. EPA, "Green Remediation and Utility-Scale Solar Development: The Aerojet Gen-

eral Corporation Superfund Site and Sacramento County, California," July 2010, 5, http://www.epa.gov/.

23. Michael Girard, sustainability director, Aerojet Rocketdyne, interview with author, April 25, 2013. All quotes of Michael Girard are from this interview.

24. EPA, "RE-Powering America's Land: Siting Renewable Energy on Potentially Contaminated Land and Mine Sites—Solar Technologies," http://www.epa.gov/oswer cpa/docs/repower_technologies_solar.pdf, accessed Feb. 12, 2015. The Solar Energy Industries Association (SEIA) estimates that a megawatt of PV can supply enough power, on average, for 164 American homes, but here I use the EPA's more conservative estimate that one megawatt of solar PV can meet the power needs of 146 average American households. Enesta Jones, EPA Office of Media Relations, e-mail to author, April 29, 2014.

25. EPA, RE-Powering America's Land Initiative: Project Tracking Matrix, November 2013, http://www.epa.gov/oswercpa/docs/tracking_matrix.pdf.

26. This survey favorably screened 320 federal facilities for PV installations greater than 300 kilowatts, and 196 federal facilities for PV installations greater than 6.5 megawatts in installed capacity. Enesta Jones, EPA Office of Media Relations, memorandum to author, April 3, 2014.

27. The White House, *Presidential Memorandum—Federal Leadership on Energy Management*, December 5, 2013, http://www.whitehouse.gov/.

28. Site assessment and cleanup grants are generally capped at $200,000 and are made available to local and state governments, counties, regional councils, planning commissions, and Indian tribes. Revolving loan funds are capitalized at the level of $600,000 to $1 million per fund and may be established through the same entities. More than 130 of these funds exist, and the EPA creates roughly ten new funds annually. Enesta Jones, EPA Office of Media Relations, memoranda to author, April 3, 2014, and April 17, 2014.

29. Adam Klinger, senior analyst, Center for Policy Analysis, Office of Solid Waste and Energy Response, USEPA, phone interview with author, April 3, 2014. See also EPA, "Harvard University Recognizes EPA Renewable Energy Program as a Top Government Innovation," news release, May 1, 2013, http://yosemite.epa.gov/.

CHAPTER FIVE: THE DESERT'S HARVEST

1. Denholm and Margolis, "Land-Use Requirements and the Per-Capita Solar Footprint for Photovoltaic Generation in the United States," 3531, 3539. Total US land area is 3,535,932 square miles, according to the US Census Bureau.

2. According to NREL, 22 percent of the total US residential roof area in cool climates and 27 percent of the residential roof area in warm/arid climates (due to reduced tree shading) is available for solar. Denholm and Margolis, *Supply Curves for Rooftop Solar PV-Generated Electricity for the United States*, 4. Actual availability is further constrained, however, by limitations facing renters and dwellers in multi-unit buildings who may not have the legal right to install solar on their roofs.

3. Denholm and Margolis, "Land-Use Requirements," 3538–39. The urban area footprint in US cities is 837 meters per person, according to this study.

4. Concentrating solar power, or CSP, is the subject of chapter 7.

5. For a more complete description of NREL's methodology, including the exclusions and constraints for different categories of solar power, see Lopez et al., *U.S. Renewable Energy Technical Potentials*, app. A.

6. See Burr et al., *Shining Cities*, app. 4, table B-1.

7. Utility-scale solar plants (both PV and concentrating solar power) brought 4,701 megawatts of new capacity online in 2014. That constituted 67.4 percent of solar capacity additions (6,968 megawatts in all) that year. SEIA/GTM Research, *2014 Year in Review*, Executive Summary, 9, 10, 19, http://www.seia.org/.

8. Roger Tobler, mayor, Boulder City, NV, interview with author, November 27, 2012.

9. Lisa Briggs, regional director, external affairs, Sempra U.S. Gas & Power, e-mail to author, February 27, 2014. See additional information on Copper Mountain Solar Project at http://www.semprausgp.com/energy-solutions/.

10. Michael Mishak, "Harry Reid Kicks Off Campaign Tour in Searchlight," *Las Vegas Sun*, April 5, 2010, http://www.lasvegassun.com/.

11. Lisa Briggs, e-mail to author, February 27, 2014.

12. The White House, Office of the Press Secretary, "Remarks by the President on Energy, Copper Mountain Solar Project, Boulder City, Nevada," March 21, 2012, http://www.whitehouse.gov/.

13. Lisa Briggs, interview with author, November 28, 2012.

14. The government's projected number of households served by 23.7 gigawatts of solar power assumes that some of this power will come from concentrating solar power (CSP) facilities, whose expected output per megawatt of installed capacity varies widely from one facility to the next but is estimated to be substantially higher than that of PV-based solar farms. For the installed capacity of major CSP plants and the number of households served by their electricity, see SEIA, "Issues & Policies: Concentrating Solar Power," http://www.seia.org/, accessed March 25, 2015.

15. US Bureau of Land Management (BLM), "Obama Administration Approves Roadmap for Utility-Scale Solar Energy Development on Public Lands," press release, October 12, 2012, http://www.doi.gov/. For detailed documentation, see BLM and Department of Energy, *EIS-0403: Final Programmatic Environmental Impact Statement, Solar Energy Development in Six Southwestern States (AS, CA, CO, NV, MN, and UT)*, July 2012, http://www.energy.gov/.

16. This facility relies on thin film solar panels rather than silicon crystal-based PV panels. The attributes of these different technologies are described in chapter 8. Thin film panels (1.2 meters by 0.60 meters) are typically smaller in size than utility-scale crystalline panels (1.6 meters by 0.80 meters), and they are also less efficient than their crystalline counterparts at converting sunlight to electricity. A greater number of them is therefore required to achieve a given power output.

17. Joseph H. Rowling, vice president, project development, Sempra, interview with author, November 28, 2012.

18. President Barack Obama, "Diversifying Our Energy Portfolio," speech at Copper Mountain Solar Projects, March 21, 2012, YouTube, http://www.youtube.com/.

19. Western Lands Project et al. v. U.S. Bureau of Land Management, Complaint for Declaratory Relief, February 12, 2013, http://www.scribd.com/doc/125184283/PEIS-Final-Complaint.

20. Chris Clarke, "Groups Sue U.S. Government over Solar Plan," February 12, 2013, KCET Rewire, http://www.kcet.org/.

21. Kim Delfino, Defenders of Wildlife, letter to Steven McMasters, project manager, County Planning & Building Department, San Luis Obispo, CA, January 3, 2011, regarding

Topaz Solar Farm Conditional Use Permit (DRC2008–00009), http://www.slocounty
.ca.gov/.

22. Nicole "Nikki" Nix, community relations manager, SunPower Corp., interview with
author, April 29, 2013. See also "Cal Poly Study: California Valley Solar Project Could Inject
Millions into Local Economy," (San Luis Obispo, CA) *Tribune*, January 12, 2011, http://
www.sanluisobispo.com/.

23. Board of Supervisors Conditions of Approval, Conditional Use Permit DRC2008–
00097 (CVSR/High Plains Ranch II, LLC), http://www.slocounty.ca.gov/; Settlement
Agreement and Release, Non-Confidential Version, August 4, 2011, http://op.bna.com
/env.nsf/id/jsun-8kmu4r/$File/Carrizo%20Settlement.pdf. See also Statement on Settle-
ment Agreement between National Environmental Organizations and Solar Development
Companies Regarding San Luis Obispo Solar Projects, August 9, 2011, http://investor
.firstsolar.com/.

24. See "28 Solar Workers Sickened by Valley Fever in San Luis Obispo County," *Los
Angeles Times*, May 1, 2013.

25. Ileene Anderson, biologist and wildlands deserts director, Center for Biological Diversity, phone interview with author, August 19, 2013.

26. Unless otherwise noted, these and other wildlife-related observations about California Valley Solar Ranch were provided by Bill Alexander, owner's representative, NRG
Energy Inc. (interview with author, April 29, 2013, and e-mails to author, August 20, 2013,
and February 13, 2014). See also California Department of Fish and Game, Central Region,
California Endangered Species Act Incidental Take Permit No. 2081–2011–044–04, California Valley Solar Ranch, http://www.slocounty.ca.gov/.

27. California Valley Solar Ranch, Highlights and Fact Sheet, http://www.california
valleysolarranch.com/, accessed January 1, 2015.

28. California Desert Conservation Act of 1994, Pub.L. 103–433. Among this act's key
provisions, it added 1.2 million acres to Death Valley National Monument and made it a
national park. It also expanded the Joshua Tree National Monument by 234,000 acres
and elevated it to national park status, and it created a 1.6-million-acre Mojave National
Preserve.

29. The California Desert Conservation Act of 1994 and later bills introduced by
Sen. Feinstein are summarized at http://www.feinstein.senate.gov/public/_named
_files/106900_desert_booklet.pdf.

30. Concentrating solar power facilities such as Ivanpah are the focus of chapter 7.

31. Joe Nelson, "California Desert Protection Act Turns 20, Celebrations Planned," *San
Bernardino Sun*, August 10, 2014, http://www.sbsun.com/.

32. Governor of California, Executive Order S-14–08, November 17, 2008, http://www
.dmg.gov/documents/EO_S_14_08_Renewable_Energy_CA_111708.pdf.

33. Michael Picker, then senior adviser to the governor of California for renewable energy facilities, interview with author, May 26, 2013. For updated information on the Desert
Renewable Energy Conservation Plan, see the plan's website at http://www.drecp.org/.

34. This appointment came in the wake of a series of scandals implicating CPUC com-
missioners and staff in ex parte communications with utility executives. The incumbent
president, Michael Peevey, announced in October 2014 that he would not seek reappoint-
ment to the commission. Bay City News, "Gov. Brown Appoints New President of CPUC,"
San Francisco Examiner, December 23, 2014, http://www.sfexaminer.com/.

CHAPTER SIX: TRIBAL SUN

1. US Department of Energy, Office of Indian Energy, *Developing Clean Energy Projects on Tribal Lands*, 3.

2. Ibid., table at 33. According to this data, there are 21.6 million gigawatt-hours of renewable energy potential on tribal lands, 94 percent of which (or about 20.3 million gigawatt-hours) comes from the sun. This is about five times total US electric generation, which amounted to 4.1 million gigawatt-hours in 2013. US Energy Information Administration, *Monthly Energy Review*, December 2014, table 7.2a, "Electricity Net Generation: Total (All Sectors)," http://www.eia.gov/.

3. "Hopi Tribal Council Bans Environmental Groups," *Navajo-Hopi Observer*, September 29, 2009, http://nhonews.com/.

4. "Navajos Seek Water Agreement for Coal-Plan Lease Renewal," *California Energy Markets*, May 3, 2013.

5. Rebecca Fairfax Clay, "Tribe at a Crossroads: The Navajo Nation Purchases a Coal Mine," *Environmental Health Perspectives* 122, no. 4 (April 2014), http://ehp.niehs.nih.gov/.

6. William Anderson, "Pollution Is a Fact of Life When Living Next to the Reid Gardner Plant," *Las Vegas Sun*, November 6, 2011, http://www.lasvegassun.com/.

7. Nevada Senate Bill 123 (2013), http://www.leg.state.nv.us/.

8. Sierra Club, "Sierra Club and Moapa Band of Paiutes Sue NV Energy for Cleanup of Contaminated Reid Gardner Site," news release, August 8, 2013, https://content.sierraclub .org/.

9. William Anderson, then tribal chair, Moapa Band of Paiutes, phone interview with author, December 18, 2012. Anderson served as tribal chair from 2010 to 2014.

10. "Obama Fast-Tracks Seven Renewable Energy Infrastructure Projects," *RenewableEnergyWorld.com*, August 17, 2012, http://www.renewableenergyworld.com/; US Department of the Interior, "Salazar Approves First-Ever Commercial Solar Energy Project on American Indian Trust Lands," news release, June 21, 2012, http://www.doi.gov/.

11. "Los Angeles Takes Major Step Toward Clean Energy Future as LADWP Approves New Solar Power Agreements," *LADWP News*, October 4, 2012; "Moapa Paiute Tribe, LADWP and First Solar Break Ground on 250 MW Solar Project," *LADWP News*, March 21, 2014, http://www.ladwpnews.com/. This project, now owned by First Solar, is expected to become operational by mid-2015. Sherryl Patterson, administrative assistant, Moapa Band of Paiutes, phone interview with author, January 6, 2015.

12. The second Moapa Paiute solar plant, advanced by RES Americas, experienced a setback in October 2014, when the Nevada Public Utilities Commission declined to approve it as part of NV Energy's submitted plan for closing the Reid Gardner Generating Station. The solar plant could reemerge, however, as part of a competitive request for proposals to supply substitute power, needed once Reid Gardner shuts down. See Sean Whaley, "Nevada PUC Rejects NV Energy Plan for Moapa Solar Plant," *Las Vegas Review-Journal*, October 27, 2014, http://www.reviewjournal.com/.

13. "Ground-Breaking Solar Agreement Between LA and Nevada Tribe," *Daily Kos*, December 6, 2012, http://dailykos.com/.

14. GRID Alternatives, "Tribal Program," http://www.gridalternatives.org/, accessed November 3, 2014.

15. Grand Canyon Trust, "Community-Based Clean Energy Initiatives," July 7, 2011, http://www.grandcanyontrust.org/.

16. John Lewis, chairman, GRICUA—Gila River Indian Community Utility Authority, interview with author, December 5, 2013.

17. Phil Bautista, director, American Indian solar development, SolarCity, phone interview with author, January 3, 2014. For updated information on the states where SolarCity operates, see the company's website at http://www.solarcity.com/.

18. For more on US government efforts in this area, see "Statement of Tracey A. LeBeau, Director, Office of Indian Energy Policy and Programs, DOE, before the Committee on Indian Affairs, US Senate, February 16, 2012," http://energy.gov/.

19. See Kirk Siegler, "Radio Station KYAY Is Lifeline for Apache Tribe," National Public Radio, September 3, 2013, http://www.npr.org/.

20. Ken Duncan, then energy coordinator, San Carlos Apache Tribe, interview with author, December 10, 2013. On the same date, financial details about the casino installation were presented at the San Carlos Tribal Energy Summit by Joe Papa, senior account executive at Ameresco, the company that has been hired to install the casino's solar power.

21. Interview with Tony Duncan of Estun-Bah, Canyon Records, May 4, 2010, YouTube, https://www.youtube.com/.

22. Winona LaDuke in *Power Paths*, documentary, Looking Hawk Productions, 2009.

23. Defenders of Wildlife, "Desert Tortoise Fact Sheet," http://www.defenders.org/.

24. Darren Daboda, Moapa Paiute tribal chair (appointed in 2014) and former environmental director, phone interview with author, January 12, 2015. A total of 128 tortoises were found in the project area, but only those inside the fenced-in solar fields were relocated.

25. Ileene Anderson, biologist and wildlands deserts director, Center for Biological Diversity, phone interview with author, August 19, 2013.

26. Figueroa, *Ancient Footprints of the Colorado River*, iv.

27. See, e.g., David Kelly, "Near Blythe, Historian Sees Solar Plants as Threat to Desert Carvings," *Los Angeles Times*, April 24, 2010, http://articles.latimes.com/.

28. In two federal lawsuits, the court dismissed La Cuna's claims that consultation with tribal groups was insufficient. La Cuna de Aztlán Sacred Sites Protection Circle v. Western Area Power Administration, 2012 WL 6743790 (C.D. Cal. November 29, 2012); La Cuna de Aztlán Sacred Sites Protection Circle v. U.S. Department of Interior, 2013 WL 4500572 (C.D. Cal. August 16, 2013). See also Phil Willon and Tiffany Hsu, "Lawsuit Alleges Solar Projects Would Harm Sacred Native American Sites," *Los Angeles Times*, February 24, 2011, http://articles.latimes.com/; Naoki Schwartz and Jason Dearen, "Native American Groups Sue to Stop Solar Projects," *Bloomberg Businessweek*, February 28, 2011, http://www.businessweek.com/.

29. Alfredo Acosta Figueroa, La Cuna de Aztlán Sacred Sites Protection Circle, phone interview with author, January 24, 2014.

30. Rebecca Tsosie, Regents' Professor of Law, Arizona State University, interview with author, December 11, 2013. See bibliography for articles authored by Tsosie.

CHAPTER SEVEN: FOCUSING ON TONOPAH

1. SEIA/GTM Research, *2014 Year in Review*, Executive Summary, 3, http://www.seia.org/.

2. For US coal plant unit sizes, see SourceWatch, "Existing Coal Plants: Size Comparison of Coal Plants," http://www.sourcewatch.org/, accessed November 3, 2014. As of that date, SourceWatch reported that there were 392 US coal generating units with 500

megawatts of capacity or less; 97 in the 500–1,000 megawatt range; and 97 with 1,000 megawatts or greater capacity.

3. According to preliminary data from the Solar Electric Power Association, utility-scale PV (defined as all systems above 5 megawatts) accounted for 9.62 to 9.96 gigawatts of installed capacity by the end of 2014. Tanuj Deora, chief strategy officer, Solar Electric Power Association (SEPA), e-mail to author, March 16, 2015.

4. US Department of Energy, "Concentrating Solar Power Dish/Engine System Basics," August 20, 2013, http://energy.gov/.

5. Bobby Jean Roberts, advertising manager, *Tonopah [Nevada] Times-Bonanza*, interview with author, November 28, 2012.

6. Bill Roberts, "Mizpah, SolarReserve Are for Real," *Tonopah Times-Bonanza*, August 25, 2011.

7. Brian Painter, Crescent Dunes site manager, SolarReserve, interview with author, November 30, 2012. Additional technical specifications were obtained from project descriptions on the SolarReserve website, http:www.solarreserve.com/, and from e-mail communications with Andi Plocek, director of communications, March 17–24, 2014.

8. The loan guarantee program and other federal renewable energy incentives are examined more closely in chapters 9 and 10.

9. References to Kevin Smith, CEO, SolarReserve, are based on phone interviews with the author conducted on October 25, 2012, and February 7, 2014.

10. Joni Eastley, Nye County commissioner (since retired), and James T. Eason, town manager, Tonopah, interviews with author, December 1, 2012.

11. US Department of the Interior, *Tonopah Solar Energy, LLC—Crescent Dunes Solar Energy Project: Draft Environmental Impact Statement* (DEIS), September 3, 2010, 3–39, http://www.blm.gov/.

12. US Department of the Interior, *Tonopah Solar Energy, LLC—Crescent Dunes Solar Energy Project: Final Environmental Impact Statement* (FEIS), November 19, 2010, 49, http://www.blm.gov.

13. The surrounding area is referred to as the Cumulative Effects Study Area. *Crescent Dunes DEIS*, September 3, 2010, 3–34.

14. The International Union for the Conservation of Nature (IUCN) lists the pale kangaroo mouse as a species of "least concern" with a decreasing population trend on its Red List of Threatened Species, http://www.iucnredlist.org/, accessed January 7, 2015.

15. Tom Seley, field manager, BLM, interview with author, November 30, 2012. See also *Crescent Dunes DEIS*, 3–34.

16. Center for Biological Diversity, Comment 23-D: Insects, Birds, Bats, and Raptors, *Crescent Dunes FEIS*, 47–48.

17. "More Avian Mortality Data Needed before Palen Decision Is Issued," *California Energy Markets*, August 15, 2014. The developers, BrightSource and Abengoa Solar, projected 1,107 to 1,469 deaths per year, while Dr. Shawn Smallwood, a biologist who is an independent environmental services consultant, gave the higher estimate.

18. Sammy Roth, "Palen Solar Project Dropped by Developers," (Palm Springs) *Desert Sun*, September 26, 2014, http://www.desertsun.com/. Uncertainties remain about this project, however. In November 2014, Abengoa announced that it had bought out BrightSource's interest in Palen and planned to return to the Energy Commission with a revised proposal for a power tower project incorporating storage capability—one of the commis-

sion's earlier recommendations. "Abengoa to Go Solo on Palen," *California Energy Markets*, November 7, 2014.

19. Wallace P. Erickson et al., *Avian Collisions with Wind Turbines: A Summary of Existing Studies and Comparisons to Other Sources of Avian Collision Mortality in the United States*, National Wind Coordinating Committee Resource Document, August 2001, 7–12, http://www.west-inc.com/. This study estimates bird deaths from collisions with buildings at 97.6 to 976 million per year.

20. Warburg, *Harvest the Wind* (2012), 121, quoting Peter Wenzel Kruse, senior vice president for group communications and investor relations, Vestas, October 8, 2009.

21. Richard Conniff, "The Evil of the Outdoor Cat," *New York Times*, March 21, 2014.

22. Jordan Kushins, "NYC Is Home to a Hyper-Real Simulation of a Massive Nevada Solar Plant," *Gizmodo India*, October 15, 2014, http://www.gizmodo.in/.

23. California Energy Commission, "Large Solar Energy Projects," September 14, 2012, http://www.energy.ca.gov/. See also San Diego Regional Renewable Energy Study Group, *Potential for Renewable Energy in the San Diego Region*, app. E: Solar Thermal—Concentrated Solar Power, 2005, http://grist.files.wordpress.com/2010/11/appendix.pdf.

24. Information on the Solana facility was obtained from the author's site visit and interview with Emiliano Garcia, plant manager, Abengoa Solar, December 9, 2013, from follow-up e-mail correspondence with Mr. Garcia, and from the company's website at http://www.abengoa.com/.

25. According to a preliminary analysis by the National Renewable Energy Laboratory, a wet-cooled CSP plant consumes 3.5 liters per kilowatt-hour over its full life cycle (including water used in equipment manufacturing). This compares with 2.9, 2.6, and 1.4 liters of water per kilowatt-hour used by a typical nuclear, coal, and combined-cycle gas plant, respectively. Switching to a dry-cooled design would reduce a CSP plant's direct operational water demands to 0.13 liters per kilowatt-hour. Burkhardt III et al., "Life Cycle Assessment," 2461–62.

26. Ibid.

27. Chris Meehan, "SolarReserve Completes Receiver at Crescent Dunes Solar Tower," *CleanEnergyAuthority.com*, April 12, 2013, http://www.cleanenergyauthority.com/, quoting Stephen Mullenix, senior vice president of operations, SolarReserve.

28. Water needed to clean mirrors at a parabolic trough CSP plant is estimated at 0.12 liters per kilowatt-hour. Burkhardt III et al., "Life Cycle Assessment," 2459.

29. Center for American Progress, "Rep. Paul Ryan 'Path to Prosperity' Anything But for Clean Energy," March 15, 2013, http://earthtechling.com/.

30. Lazard, *Lazard's Levelized Cost of Energy Analysis—Version 8.0*, September 2014, http://www.lazard.com/.

31. Fraunhofer ISE, *Levelized Cost of Electricity—Renewable Energy Technologies Study*, 29. At direct normal solar irradiance (DNI) of 2,000 kilowatt-hours per square meter per year, Fraunhofer estimates the cost of CSP tower technology at 25.0 to 28.5 cents per kilowatt-hour and parabolic trough-based CSP at 21.8 to 26.8 cents per kilowatt-hour, with 8 hours of thermal storage capacity in both cases. Costs stated here have been converted by the author from Euros at the rate applicable on November 30, 2013 (1.3588 US$/Euro).

32. CPUC Public Agenda 3303, October 25, 2012, http://www.cpuc.ca.gov/.

33. "BrightSource Mothballs Rio Mesa Solar Thermal Project, Refocuses on Pipeline," *RenewableEnergyWorld.com*, January 23, 2013, http://www.renewableenergyworld.com/;

Chris Meehan, "BrightSource Abandons 500MW Rio Mesa Solar Project," July 15, 2013, http://www.cleanenergyauthority.com/.

34. By the end of 2014, cumulative installed solar capacity was about 20.5 gigawatts. SEIA/GTM Research, *2014 Year in Review*, Executive Summary, 3, http://www.seia.org/. Utility-scale PV capacity as of that date amounted to 9.621 to 9.965 gigawatts (direct current). Tanuj Deora, chief strategy officer, Solar Electric Power Association (SEPA), e-mail to author, March 16, 2015. At 2.2 gigawatts, CSP accounted for 10.7 percent of cumulative solar capacity (residential, non-residential, utility-scale) and 20 to 23 percent of cumulative utility-scale capacity installed through 2014.

35. Moomaw et al., *Annex II: Methodology*, table A.II.4, in IPCC *Special Report on Renewable Energy Sources and Climate Change Mitigation*.

CHAPTER EIGHT: CRADLE TO GRAVE

1. At the time of the author's tour of the Hillsboro site on December 12, 2012, all stages of PV panel production were still being carried out, but since then the company has curtailed some of its operations due to mounting overseas competition. For an exploration of the impact of global trade on SolarWorld and other players in the US solar industry, see chapter 9.

2. According to Christian Honeker, materials scientist at the Fraunhofer Center for Sustainable Energy Systems, quartz is superior to graphite for this application. Though very heat-tolerant, graphite at high temperatures leaches contaminants that can impair the quality of polysilicon. E-mail to author, June 26, 2014.

3. Ben Santarris, head of corporate communications and sustainability, SolarWorld, December 12, 2012. Except as otherwise indicated, observations about SolarWorld's manufacturing process come from this interview and site visit.

4. Jordan and Kurtz, "Photovoltaic Degradation Rates." Looking at nearly two thousand degradation rates, this study found a median value of 0.5 percent per year for both individual panels and complete solar systems.

5. Wetzel and Feuerstein, *Update of Energy Payback Time for Crystalline PV Modules*, 3193.

6. De Wild-Scholten, "Energy Payback Time and Carbon Footprint of Commercial Photovoltaic Systems." The energy calculations in this study include all stages of manufacturing but do not include installation, operation and maintenance, and end-of-life decommissioning.

7. See Terakawa, "Review of Thin-Film Silicon Deposition Techniques."

8. According to one study of polysilicon-based thin-film cells, the amount of silicon needed for these cells is lower by a factor of one hundred than the material required for crystalline-based solar cells. Becker et al., "Polycrystalline Silicon Thin-Film Solar Cells."

9. Industry analyst Dan Chiras cautions that, to achieve a cell temperature of 77°F in full sunlight, the ambient temperature must be about 23° to 32°F, "not a typical temperature for PV modules in most locations most of the year." Chiras, *Power from the Sun*, 4–5. Silicon-based panels have temperature coefficients of −0.4 to −0.5 percent per degree Celsius. Applying this coefficient, reducing a PV cell's surface temperature from 77°F (25°C) to 41–49°F (5–9°C) will achieve an output gain of 7 to 9 percent. Dropping to a truly frigid temperature of −36 to −40°F (−39 to −40°C) will result in an efficiency gain of 27 to 29

percent. Christian Honeker, materials scientist, Fraunhofer Center for Sustainable Energy Systems, e-mail to author, March 18, 2015.

10. Shyam Mehta, solar analyst, GTM Research, "Global PV Supply Trends: Pricing, Technology, Production and Profitability," Greentech Media webinar, May 8, 2014, reporting a mean conversion efficiency of 17.2 to 18.8 percent for crystalline silicon cells.

11. Ingrid Ekstrom, corporate communications director, SunPower, e-mail to author, December 30, 2013.

12. One of SunPower's larger utility-scale projects is the California Valley Solar Ranch, described in chapter 5.

13. Eric Wesoff, "Solar Frontier Grabs Thin-Film Efficiency Conversion Record from First Solar," *GreenTechSolar*, April 2, 2014, http://www.greentechmedia.com/.

14. Junko Movellan, "Bifacial PV Modules: Can They Move Beyond BIPV Applications?," *RenewableEnergyWorld.com*, January 9, 2014, http://www.renewableenergyworld.com/. Not surprisingly, bifacial modules are most productive when indirect light is greatest—for example, when the ground is snow-covered or a roof has a white or near-white surface. At one test site in Japan, bifacial modules mounted above a bed of finely crushed scallop shells were found to generate 13.5 percent more power, on average, than those mounted on a grassy surface and 23.6 percent more power than one-sided modules.

15. See Kim et al., "Remarkable Progress in Thin-Film Silicon Solar Cells Using High-Efficiency Triple-Junction Technology."

16. Eric Wesoff, "Sharp Hits 44.4% Efficiency for Triple-Junction Solar Cell," *GreenTechSolar*, June 24, 2013, http://www.greentechmedia.com/. See also Luque, "Will We Exceed 50% Efficiency in Photovoltaics?," and Guter et al., "Toward the Industrialization of Concentrator Solar Cells with Efficiencies above 40%."

17. See Badawy, "A Review on Solar Cells from Si-Single Crystals to Porous Materials and Quantum Dots."

18. Christian Honeker, materials scientist, Fraunhofer Center for Sustainable Energy Systems, interview with author, November 4, 2013.

19. Geoffrey S. Kinsey, Christian Honeker, and Cordula V. Schmid, Fraunhofer Center for Sustainable Energy Systems, interview with author, November 4, 2013. Fraunhofer's testing methodology is described in Meakin et al., "Fraunhofer PV Durability Initiative for Solar Modules."

20. For comparative data on solar deployment by states, see fig. 4 in chapter 1.

21. "Universal waste" is often applied to widely used products that tend to, but should not, end up in conventional landfills, such as lead-acid batteries, aerosol cans, discarded electronic devices, and mercury-containing fluorescent light bulbs.

22. California Dept. of Toxic Substances Control, "Solar Panel Draft Reg Text for July 2010 Workshop: Proposed Standards for Management of Waste Solar Panels," Ref. # R-2010–01, https://dtsc.ca.gov/LawsRegsPolicies/Regs/upload/Solar-Panel-Draft-Reg-Text-for-July-Workshop.pdf.

23. Sukhwant Raju, global director of recycling, operations, and services, First Solar, PowerPoint presentation: "First Solar's Industry-Leading PV Technology and Recycling Program," Solar Power International, Chicago, October 22, 2013.

24. Alex Heard, "First Solar Fires Back on Recycling Charges and Decommissioning Costs," *GreenTechSolar*, editorial, August 28, 2013, http://www.greentechmedia.com/.

25. Sheila Davis, executive director, Silicon Valley Toxics Coalition, interview with author, April 23, 2013.

26. Directive 2012/19/EU on Waste Electrical and Electronic Equipment (WEEE), July 4, 2012, Annex II and V, http://ec.europa.eu/environment/waste/weee/legis_en.htm/.

27. See PV Cycle's website, http://www.pvcycle.org/, for a description of national programs including PV Cycle UK. PV Cycle programs fall short of being fully prepaid waste recovery programs, as they only cover panels discarded during a specified membership period, rather than over the full life cycle of the panels. Dustin Mulvaney, assistant professor, environmental studies, San Jose State University, e-mail to author, June 17, 2014.

28. There is a growing trend toward frameless panels using double panes of glass or other substrates for structural support, particularly in the thin-film market. Most panels marketed today still have metal frames, however.

29. Dustin Mulvaney, phone interview with author, September 11, 2013.

30. Moomaw et al., *Annex II: Methodology*, table A.II.4, in *IPCC Special Report on Renewable Energy Sources and Climate Change Mitigation*. The cited data are median values based on aggregated results of a broad literature survey.

31. According to one study, silicon PV modules made in China consume 28–48 percent more energy than those manufactured in Europe, and greenhouse gas emissions are twice as high. Yue et al., "Domestic and Overseas Manufacturing Scenarios of Silicon-Based Photovoltaics: Life Cycle Energy and Environmental Comparative Analysis."

CHAPTER NINE: BUILDING A ROBUST SOLAR ECONOMY

1. Romney's aid package to Evergreen included a $10 million grant, a ninety-nine-year lease on the land for its factory for $1, and a $7.5 million investment tax credit. Of the Commonwealth's $23 million investment, about $11 million had been recovered as of December 2014. Jake Lambert, chief of staff, Massachusetts Executive Office of Housing and Economic Development, e-mail to author, December 19, 2014.

2. "Romney Crashes Solyndra with Press Conference," May 31, 2012, YouTube, http://www.youtube.com/.

3. Stephen Lacey and Richard Caperton, "Exclusive Timeline: Bush Administration Advanced Solyndra Loan Guarantee for Two Years, Media Blow the Story," *Climateprogress*, October 25, 2012, http://thinkprogress.org/.

4. Matthew L. Wald, "Solar Firm Aided by U.S. Shuts Doors," *New York Times*, September 1, 2011. For a longer-term representation of the decline in PV panel prices since 1975, see fig. 7 (on the following page).

5. Todd Woody, "Texas Oil Baron Is Promoting Solar Energy," *New York Times*, July 13, 2010.

6. "SolarWorld Television Advert 'Made in Germany' (Starring Larry Hagman)," published on YouTube, August 29, 2012, https://www.youtube.com/.

7. Alexander S. Fader, *Fighting Fair: A Legal Analysis of SolarWorld's Trade Dispute with China*, III(A), citing Petition for the Imposition of Antidumping and Countervailing Duties Pursuant to Sections 701 and 71 of the Tariff Act of 1930, As Amended, *In the Matter of Crystalline Silicon Photovoltaic Cells and Modules from China*, USITC Investigation Nos. 701-TA-481 and 731-TA-1190, 28.

8. Shyam Mehta, "GTM Research: Yingli Gains Crown as Top Producer in a 36 GW

Fig. 7. *Photovoltaic Panels—Price per Watt (1975–2014)*

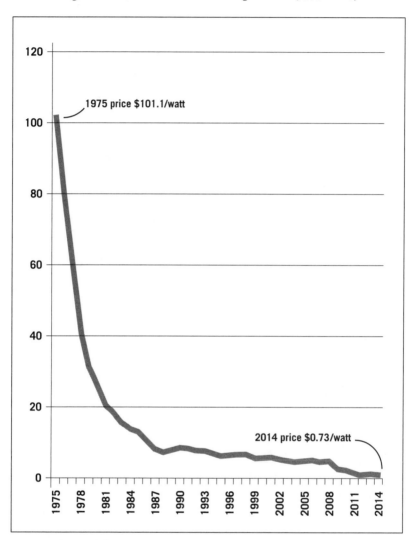

Sources: Paul Maycock (1975–2003 prices); Bloomberg New Energy Finance (2004–2013 prices), in 2013 US$. The quoted price for 2014 pertains to the final quarter of that year, as reported in SEIA/ GTM Research, *U.S. Solar Market Insight Report—2014 Year in Review.*

Global PV Market," *GreenTechSolar*, May 1, 2013, http://www.greentechmedia.com/. These percentages represent megawatts of rated solar output, not the number of PV panels that were sold.

9. Feldman et al., *Photovoltaic (PV) Pricing Trends: Historical, Recent, and Near-Term Projections*, fig. 2. Global prices cited here are based on Navigant Consulting's module price index for large-quantity buyers.

10. Asbeck, *A Solar World*, 161–67.

11. SolarWorld claimed that information about its technology development, production costs, cash flow, and legal strategies had been compromised by this infiltration. Keith Bradsher, "Retaliatory Attacks, Online: U.S. Companies That Challenge China on Trade Face Cybertheft," *New York Times*, May 21, 2014.

12. The steps in the US government's investigations and the specific roles of the Department of Commerce and the International Trade Commission are summarized in Fader, "Fighting Fair." See note 7 above.

13. "It Ain't Over Till It's Over," *Photon International*, December 2012, 32.

14. Frank Asbeck, letter to President Barack Obama, February 5, 2014, http://www.solarworld-usa.com/; Ben Santarris, director of corporate communications and sustainability, SolarWorld, phone interview with author, October 7, 2014.

15. Richard Read, "SolarWorld to Lay Off 100 in Hillsboro as Chinese Manufacturers Continue Cutting Prices," *Oregonian*, July 8, 2013, http://www.oregon.live.com/.

16. On December 16, 2014, the Commerce Department announced new antidumping tariffs ranging from 26.71 to 78.42 percent and countervailing duties, targeting Chinese subsidies, of 27.64 to 49.79 percent. Actual sanctions will be based on findings of material injury through further administrative proceedings. Julia Pyper, "New Tariffs on Chinese Solar-Panel Makers Split the US Solar Industry," *GreenTechSolar*, December 17, 2014, http://www.greentechmedia.com/.

17. Xie Yu, "Solar Firms Face 'Total Eclipse' in the US," *China Daily*, January 8, 2014, http://www.chinadaily.com/.

18. Asbeck, *A Solar World*, 167.

19. Jigar Shah, letter to the president, February 24, 2014, http://www.affordable solarusa.org/.

20. Keith Bradsher, "U.S. Solar Panel Makers Say China Violated Trade Rules," *New York Times*, October 19, 2011.

21. SEIA, "Commerce Solar Trade Decision Adds Urgency to Industry Discussions," press release, June 3, 2014, http://www.seia.org.

22. William P. Hirshman, "Off Duty: The European Commission Lifts Threat of Import Duties on Chinese Module Makers That Agree to a Minimum Price and Annual Cap," *Photon*, September 2013.

23. John Smirnow, vice president for trade and competitiveness, SEIA, phone interview with author, July 3, 2014. See also GTM Research, *The EU-China Deal: What We Know and Don't Know*, white paper, August 2013, http://forms.greentechmedia.com/.

24. Smirnow interview, July 3, 2014.

25. Shyam Mehta and Shayle Kann, *The 2014 U.S.–China Solar Trade Dispute*.

26. Shayle Kann, senior vice president, research, GTM Research, phone interview with author, August 1, 2014.

27. SunPower, *Annual Report*, 2012, http://us.sunpower.com/, 9–10.

28. Tom Werner, CEO, SunPower, phone interview with author, September 26, 2013.

29. Ingrid Ekstrom, director of corporate communications, SunPower, e-mail to author, December 30, 2013.

30. By 2014 First Solar claimed to have manufactured more than 100 million thin-film panels and installed over 8 gigawatts of solar power, relying on a supply chain employing forty thousand people globally. "Taking Energy Forward," http://www.firstsolar.com/, accessed August 6, 2014.

31. James Montgomery, "SunEdison Mulls Massive PV Factory in Saudi Arabia," *RenewableEnergyWorld.com*, http://www.renewableenergyworld.com/, February 6, 2014.

32. Shayle Kann interview, August 1, 2014.

33. Silevo, "Greater Than 21% Efficiency on Full Size Substrates and Production Proven Materials," http://silevosolar.com/, accessed October 28, 2014.

34. Ehren Goossens, "SolarCity Copies Musk's 'Gigafactory' Manufacturing Model," *Bloomberg Businessweek*, June 18, 2014, http://www.bloomberg.com/.

35. See fig. 3 in the introduction. As this figure reveals, panels comprise a larger percentage of total system price for utility-scale installations because of the lower cost per watt of other inputs such as other hardware (e.g., inverters, wiring, mounting systems), labor, engineering, and permitting. Non-residential rooftop systems occupy a middle ground in this respect.

36. J. Seel et al., "Why Are Residential PV Prices in Germany So Much Lower Than in the United States?—A Scoping Analysis," US Department of Energy SunShot, February 2013 revision, slide #46, http://emp.lbl/sites/allfiles/german-us-pv-price-ppt.pdf.

37. "Germany's Solar Power Systems Set New Solar Record (Germany Crushing US in Solar Power)," *Cost of Solar*, July 2013, http://costofsolar.com/germany-solar-power-systems/.

38. Solar Foundation, *National Solar Jobs Census 2014*, 1, http://thesolarfoundation.org/.

39. SEIA/GTM Research, *U.S. Solar Market Insight Report—Q1 2014—Full Report*, figs. 2.44, 2.46, and 2.48; *U.S. Solar Market Insight Report—Q3 2014—Executive Summary*, figs. 2–5, 2–6, 2–7. In 2014 SEIA and GTM's methodology for determining average prices shifted from quoted prices, sometimes reflecting noncurrent values, to actual prices paid for delivered systems. These two methodologies are described in the *Market Insight Report—Q1 2014* at 56–57.

40. Lazard, *Lazard's Levelized Cost of Energy Analysis*, September 2014, http://www.lazard.com/. Lazard cites these costs in dollars per megawatt-hour, which I have converted to cents per kilowatt-hour for ease of comparison with other sections of this book. It is important to note that the comparisons in this study are based on the *unsubsidized* cost of producing power from different energy sources. This makes solar PV's competitive pricing that much more impressive, as the stated costs do not factor in the 30 percent federal investment tax credit or other renewable energy subsidies discussed in chapter 10.

41. US Energy Information Administration (EIA), "Project Sponsors Are Seeking Federal Approval to Export Domestic Natural Gas," April 24, 2012, http://www.eia.gov/.

42. EIA, Henry Hub Natural Gas Spot Price, http://www.eia.gov/, released March 11, 2015. The graph and chart on this webpage reveal the extreme price volatility of natural gas pricing since 1997.

CHAPTER TEN: DISRUPTING THE UTILITY STATUS QUO

1. For an examination of federal energy subsidies, see Pfund and Healey, *What Would Jefferson Do?*

2. Energy Policy Act of 2005, P.L. 109–58, 26 USC Sec. 25D (residential) and Sec. 48 (commercial).

3. Emergency Economic Stabilization Act of 2008, Division B—Energy Improvement and Extension Act of 2008, P.L. 110–343, Sec. 103. See also Baker Tilly, "Investment Tax Credit—Section 48," http://bakertilly.com/, accessed August 29, 2014.

4. American Recovery and Reinvestment Tax Act of 2009, P.L. 111–5, Sec. 1603.

5. US Department of Treasury, "Overview and Status Update of the Section 1603 Program," May 13, 2014, http://www.treasury.gov/. Applicants had to apply for this grant before October 1, 2012, and had to certify that construction of the project had already begun by that date.

6. Tom Kimbis, vice president of executive affairs, SEIA, e-mail to author, June 26, 2014. See also SEIA, "1603 Issues & Policies: Treasury Program," http://www.seia.org/.

7. Chris Meehan, "US DOE Loan Program Financed $16 Billion in Renewables and More Coming," *RenewableEnergyWorld.com*, August 28, 2013, http://www.renewable energyworld.com/.

8. "Written Statement of Peter Davidson, Executive Director, Loans Program Office, U.S. Department of Energy, Before the Senate Energy and Natural Resources Committee, July 18, 2013," http://www.energy.senate.gov/.

9. US Department of Labor, "American Recovery and Reinvestment Act of 2009: Energy Training Partnership Grants," http://www.doleta.gov/, accessed August 24, 2014.

10. HUD, "Green Retrofit Program for Multifamily Housing," http://portal.hud.gov/, accessed August 24, 2014.

11. DSIRE, "Renewable Portfolio Standard—New York," updated February 4, 2015, http://programs.dsireusa.org/.

12. For ratings of the forty-six statewide net metering policies now in place, see "Freeing the Grid 2015: Best Practices in State Net Metering Policies and Interconnection Procedures," http://freeingthegrid.org/.

13. Bart Barnes, "Barry Goldwater, GOP Hero, Dies," *Washington Post*, May 30, 1998, http://www.washingtonpost.com/.

14. TUSK advertisement, "It's your right to go solar," http://sonoranalliance.com/2013 /05/24/tusk-launches-new-ad-against-800-lb-utility-monopoly/, accessed January 6, 2015.

15. DSIRE, "Renewables Portfolio Standard–Arizona," http://www.dsire.org/, updated September 5, 2014.

16. Greg Bernosky, manager of state regulation, APS, interview with author, December 3, 2013.

17. Ryan Randazzo and Robert Angien, "APS, Solar Companies Clash over Credits to Customers," *Arizona Republic*, October 21, 2013, http://www.arizonacentral.com/.

18. APS, *Application to the Arizona Corporation Commission, In The Matter of the Application of Arizona Public Service Company for Approval of Net metering Cost Shift Solution*, Docket No. E-01345A-13–0248, July 12, 2013, 9, https://edocket.azcc.gov/.

19. Beach and McGuire, *The Benefits and Costs of Solar Distributed Generation for Arizona Public Service*, table 1 (showing total ratepayer benefits at 21.5 to 23.7 cents per kilowatt-hour and costs ranging from 13.9 to 15.5 cents per kilowatt-hour).

20. Ryan Randazzo, "ACC OKs New Solar Fees," *Arizona Republic*, November 15, 2013, http://www.arizonacentral.com/.

21. "Solar Net Metering: Arizona Led the Way," transcript, *Arizona Republic*, January 10, 2014, http://www.arizonacentral.com/.

22. Cathy Proctor, "Colorado's Solar-Power Goal: Grow It Tenfold," *Denver Business Journal*, January 16, 2013, http://www.bizjournals.com/.

23. California Assembly Bill 327 (signed by governor on October 7, 2013).

24. David Crane, CEO, NRG Energy, Keynote Fireside Chat, 2013 MIT Energy Conference, March 1, 2013, http://video.mit.edu/.

25. The top five US residential solar developers in 2014 were SolarCity (34 percent), Vivint (13 percent), Sungevity (3 percent), Sunrun (2 percent), and Verengo (2 percent). GTM Research, *US PV Leaderboard – Q1 2015*, http://www.greentechmedia.com.

26. Diane Cardwell, "NRG Energy Buys Goal Zero, a Start-Up, as Entry to Mobile Solar Business," *New York Times*, August 15, 2014.

27. Kind, *Disruptive Challenges*, 3.

28. Ibid., 18.

29. Christoph Steitz, "German Utilities Eye Solar Leasing to Help Turnaround," Reuters, September 3, 2014, http://www.reuters.com/.

30. Andy Colthorpe, "Utilities and Solar Groups Both Claiming Victory in Arizona Rooftop Showdown," *PV TECH*, December 22, 2014, http://www.pv-tech.org/. The Arizona Corporation Commission allowed these two utilities to proceed with solar leasing on a pilot basis, charging residential customers a fixed fee over a twenty-five-year period.

31. For a helpful overview, see US Department of Energy, *A Guide to Community Shared Solar*.

EPILOGUE: OUR SOLAR FUTURE

1. Jess Bidgood, "Charges Dropped Against Climate Activists," *New York Times*, September 8, 2014.

2. Lopez et al., *U.S. Renewable Energy Technical Potentials*, table 12. As indicated earlier, this table estimates that 800 terawatt-hours of electricity per year could come from rooftop photovoltaics, compared to total US retail electricity sales of 3,754 terawatt-hours in 2010.

3. See IPCC *Fifth Assessment Synthesis Report, Approved Summary for Policymakers*, November 1, 2014, http://www.ipcc.ch/pdf/assessment-report/ar5/syr/SYR_AR5_SPM.pdf.

4. European Commission, "Climate Action: EU Greenhouse Gas Emissions and Targets," http://ec.europa.eu/clima/policies/g-gas/index_en.htm, updated July 9, 2014.

5. European Commission, "Climate Action: EU Action on Climate," http://ec.europa.eu/clima/policies/brief/eu/index_en.htm, updated September 5, 2014.

6. Danish Government, *Climate Policy Plan*, 14.

7. In 2014, officially reported as a "normal wind year," 39 percent of Denmark's electricity consumption came from wind. "Denmark Nearing 2020 Wind Energy Target," *Copenhagen Post*, January 6, 2015, http://cphpost.dk/, citing Energinet.dk, a nonprofit enterprise owned by the Danish Climate and Energy Ministry.

8. Danish Government, *Climate Policy Plan*, 10.

9. Justin Gillis, "Sun and Wind Alter Global Landscape, Leaving Utilities Behind," *New York Times*, September 13, 2014.

Selected Bibliography

ACORE, *US Department of Defense & Renewable Energy: An Industry Helping the Military Meet Its Strategic Energy Objectives.* January 2012. http://www.acore.org/.

Alic, John. *Energy Innovation at the Department of Defense: Assessing the Opportunities.* Clean Air Task Force and Consortium for Science, Policy & Outcomes, March 2012.

Asbeck, Frank. *A Solar World: The SolarWorld CEO on the Future of Our Energy Supply.* Cologne: Kiepenheuer & Witsch, 2009, 2012.

Badawy, Waheed A. "A Review on Solar Cells from Si-Single Crystals to Porous Materials and Quantum Dots." *Journal of Advanced Research* (2013).

Barbose, Galen, Naïm Darghouth, Samantha Weaver, and Ryan Wiser. *Tracking the Sun VI: An Historical Summary of the Installed Price of Photovoltaics in the United States from 1998 to 2012.* US Department of Energy, July 2014. http://emp.lbl.gov/.

Barbose, Galen, Samantha Weaver, and Naïm Darghouth. *Tracking the Sun VII: An Historical Summary of the Installed Price of Photovoltaics in the United States from 1998 to 2013.* US Department of Energy, September 2014. http://emp.lbl.gov/.

Beach, R. Thomas, and Patrick G. McGuire. *The Benefits and Costs of Solar Distributed Generation for Arizona Public Service.* Crossborder Energy, May 8, 2013. http://www.seia.org/.

Becker, C., D. Amkreutz, T. Sonheimer, V. Preidel, et al. "Polycrystalline Silicon Thin-Film Solar Cells: Status and Perspectives." *Solar Energy Materials & Solar Cells* 119 (2013): 112–23.

Bolinger, Mark, and Samantha Weaver. *Utility-Scale Solar 2013: An Empirical Analysis of Project Cost, Performance, and Pricing Trends in the United States.* Lawrence Berkeley Laboratory, September 2014. http://emp.lbl.gov/.

Burkhardt, Jesse, Ryan Wiser, Naïm Darghouth, C. G. Dong, et al. *How Much Do Local Regulations Matter? Exploring the Impact of Permitting and Local Regulatory Processes on PV Prices in the United States.* Lawrence Berkeley Laboratory, September 2014. http://emp.lbl.gov/.

Burkhardt, John J., III, Garvin A. Health, and Craig S. Turchi. "Life Cycle Assessment of a Parabolic Trough Concentrating Solar Power Plant and the Impacts of Key Design Alternatives." *Environmental Science & Technology* 45 (2011): 2457–64.

Burr, Judee, Tony Dutzik, Jordan Schneider, and Rob Sargent. *Shining Cities: At the Forefront of America's Solar Energy Revolution.* Environment Massachusetts, April 2014. http://www.environmentamerica.org/.

Burr, Judee, Lindsey Hallock, and Rob Sargent. *Star Power: The Growing Role of Solar Energy in America,* November 2014. http://www.environmentamerica.org/.

California Energy Commission. *Estimated Cost of New Renewable and Fossil Generation in California.* Draft Staff Report, CEC-200–2014–003-SD, May 2014. http://www .energy.ca.gov/.

Campbell, Matt. The Drivers of the Levelized Cost of Electricity for Utility-Scale Photovoltaics. SunPower, 2009. http://large.stanford.edu/courses/2010/ph240/vasudev1 /docs/sunpower.pdf.

Carle, Jon E., and Frederik C. Krebs. "Technological Status of Organic Photovoltaics (OPV)." *Solar Energy Materials & Solar Cells* 119 (2013): 309–10.

Chiras, Dan. *Power from the Sun: Achieving Energy Independence.* Gabriola Island, BC: New Society, 2009.

Churchill, Ward. *Struggle for the Land: Native North American Resistance to Genocide, Ecocide and Colonization.* San Francisco: City Lights, 2002.

Cohen, Maurie J., Halina Szejnwald Brown, and Philip J. Vergragt. *Innovations in Sustainable Consumption: New Economics, Socio-Technical Transitions and Social Practices.* Cheltenham, UK, and Northampton, MA: Edward Elgar, 2013.

Dale, Michael, and Sally M. Benson. "Energy Balance of the Global Photovoltaic (PV) Industry—Is the PV Industry a Net Electricity Producer?" *Environmental Science & Technology* 47 (2013): 3482–89.

Danish Government. *Climate Policy Plan: Towards a Low Carbon Society.* August 2013. http://www.ens.dk/sites/ens.dk/files/policy/danish-climate-energy-policy/danish climatepolicyplan_uk.pdf.

Dastrup, Samuel, Joshua S. Graff Zivin, Dora Costa, and Matthew E. Kahn. *Understanding the Solar Home Price Premium: Electricity Generation and "Green" Social Status.* National Bureau of Economic Research, Working Paper No. 17200, July 2011. http://www .nber.org/.

De Guow, J. A., D. D. Parrish, G. J. Frost, and M. Trainer. "Reduced Emissions of CO_2, NOx, and SO_2 from U.S. Power Plants Owing to Switch from Coal to Natural Gas with Combined Cycle Technology." *Earth's Future* 2 (February 2014): 75–82, doi:10.1002/2013 EF000196. Published online February 21, 2014. http://onlinelibrary.wiley.com/.

Denholm, Paul, and Robert M. Margolis. "Land-Use Requirements and the Per-Capita Solar Footprint for Photovoltaic Generation in the United States." *Energy Policy* 36 (2008): 3531–43.

———. *Supply Curves for Rooftop Solar PV-Generated Electricity for the United States.* National Renewable Energy Laboratory, NREL/TP-6A0–44073. http://www.nrel.gov /docs/.

de Wild-Scholten, M. J. "Energy Payback Time and Carbon Footprint of Commercial Photovoltaic Systems." *Solar Energy Materials and Solar Cells* 119 (2013): 296–305.

El Chaar, L., L. A. Lamont, and N. El Zein. "Review of Photovoltaic Technologies." *Renewable and Sustainable Energy Reviews* 15 (2011): 2165–75.

Environment Massachusetts. *Shining Cities: At the Forefront of America's Solar Energy Revolution,* April 2014. http://environmentmassachusettscenter.org/.

European Association for the Recovery of Photovoltaic Modules. *Annual Report 2012.* http://www.pvcycle.org/.

European Parliament and European Council. Directive 2012/19/EU of 4 July 2012 on Waste Electrical and Electronic Equipment (WEEE) (recast). *Official Journal of the European Union* (July 24, 2012). http://ec.europa.eu/.

European Photovoltaic Industry Association. *Global Market Outlook for Photovoltaics 2014–2018*, 2014. http://www.epia.org/.

Fader, Alexander S. *Fighting Fair: A Legal Analysis of SolarWorld's Trade Dispute with China*. The Butler Firm, PLLC, 2012. http://www.thebutlerfirm.com/.

Feldman, David, Galen Barbose, Robert Margolis, Ryan Wiser, et al. *Photovoltaic (PV) Pricing Trends: Historical, Recent, and Near-Term Projections*. US Department of Energy, DOE/GO-102012–3839, November 2012. http://www.nrel.gov/docs/.

Figueroa, Alfredo A. *Ancient Footprints of the Colorado River*. Blythe, CA: La Cuna de Aztlán Publishing, 2012.

Fraunhofer ISE. *Levelized Cost of Electricity—Renewable Energy Technologies Study*. November 2013. http://www.ise.fraunhofer.de/.

Fthenakis, Vasilis M., Hyung Chul Kim, and Erik Alsema. "Emissions from Photovoltaic Life Cycles." *Environmental Science & Technology* 42, no. 6 (2008): 2168–74.

Gedicks, Al. *The New Resource Wars: Native and Environmental Struggles Against Multinational Corporations*. Boston: South End, 1993.

Glennon, Robert, and Andrew W. Reeves. "Solar Energy's Cloudy Future." *Arizona Journal of Environmental Law and Policy* 1, no. 1 (Fall 2010). http://www.ajelp.com/.

Goad, Jessica, Christy Goldfuss, and Tom Kenworthy. *Using Public Lands for the Public Good: Rebalancing Coal and Renewable Electricity with a Clean Resources Standard*. Center for American Progress, June 2012. http://www.americanprogress.org/.

Golnas, A. "PV System Reliability: An Operator's Perspective." *IEEE Journal of Photovoltaics* 3, no. 1 (January 2013): 416–21.

Guha, Subhendu, Jeffrey Yang, and Baojie Yan. "High Efficiency Multi-Junction Thin Film Silicon Cells Incorporating Nanocrystalline Silicon." *Solar Energy Materials & Solar Cells* 119 (2013): 1–11.

Guter, W., M. Meusel, W. Köstler, R. Kern, et al. "Toward the Industrialization of Concentrator Solar Cells with Efficiencies above 40%," in A. W. Brett, F. Dimroth, R. D. McConnell, and G. Sala, eds. *6th International Conference on Concentrating Photovoltaic Systems*. American Institute of Physics, 2010.

Halasah, Suleiman A., David Pearlmutter, and Daniel Feuermann. "Field Installation versus Local Integration of Photovoltaic Systems and Their Effect on Energy Evaluation Metrics." *Energy Policy* 52 (2013): 462–71.

Hand, M. M., S. Baldwin, E. DeMeo, J. M. Reilly, et al., eds. *Renewable Electricity Futures Study*. National Renewable Energy Laboratory, NREL/TP-6A20–52409, 2012. http://www.nrel.gov/docs/.

Hayes, Denis. *Rays of Hope: The Transition to a Post-Petroleum World*. New York: W. W. Norton, 1977.

Hering, Garrett. "Losing Steam? Siemens Exits Solar After CSP Catastrophe, As Solar Thermal Industry Confronts High Cost and Tempered Market Outlook." *Photon International* (December 2012): 74–80.

Hoen, Ben, Sandra Adomatis, Thomas Jackson, Joshua Graff-Zivin, et al. *Selling into the Sun: Price Premium Analysis of a Multi-State Dataset of Solar Homes*. Lawrence Berkeley National Laboratory, January 2015. http://emp.lbl.gov/publications/.

Hurlbut, David, Joyce McLaren, and Rachel Gelman. *Beyond Renewable Portfolio Standards: An Assessment of Regional Supply and Demand Conditions Affecting the Future of Renewable Energy in the West.* National Renewable Energy Laboratory NREL/TP-6A20–57830, August 2013. http://www.nrel.gov/docs/.

International Energy Agency. *PVPS Report—Snapshot of Global PV 1992–2013.* Report IEA-PVPS T1–24:2014. http://www.eia-pvps.org/.

Interstate Renewable Energy Council (IREC). *Freeing the Grid 2013: Best Practices in State Net Metering Policies and Interconnection Procedures.* October 2013. http://freeing thegrid.org/.

———. *Model Rules for Shared Renewable Energy Programs.* 2013. http://www.irecusa .org/.

———. *A Regulator's Guidebook: Calculating the Benefits and Costs of Distributed Solar Generation.* October 2013. http://www.irecusa.org/.

IRENA. *Renewable Energy Prospects: United States of America, Remap 2030 Analysis.* Abu Dhabi, 2015. http://www.irena.org/remap/.

Johnstone, Bob. *Switching to Solar: What We Can Learn from Germany's Success in Harnessing Clean Energy.* New York: Prometheus, 2011.

Jordan, D. C., and S. R. Kurtz. "Photovoltaic Degradation Rates—An Analytical Review." *Progress in Photovoltaics: Research and Applications* 21 (2013): 12–29. http://online library.wiley.com/.

Keane, Brian F. *Green Is Good: Save Money, Make Money, and Help Your Community Profit from Clean Energy.* Guilford, CT: Lyons, 2012.

Kennedy, Danny. *Rooftop Revolution: How Solar Power Can Save Our Economy—and Our Planet—from Dirty Energy.* San Francisco: Berrett-Koehler, 2012.

Kiatreungwattana, Kosol, Gail Mosey, Shea Jones-Johnson, et al. *Best Practices for Siting Solar Photovoltaics on Municipal Solid Waste Landfills.* National Renewable Energy Laboratory, NREL/TP-7A30–52615, February 2013. http://www.nrel.gov/docs/.

Kim, Soohyun, Jin-Won Chung, Hyun Lee, Jinhee Park, et al. "Remarkable Progress in Thin-Film Silicon Solar Cells Using High-Efficiency Triple-Junction Technology." *Solar Energy Materials & Solar Cells* 119 (2013): 26–35.

Kind, Peter. *Disruptive Challenges: Financial Implications and Strategic Responses to a Changing Retail Electric Business.* Edison Electric Institute, January 2013. http://www.eei .org/.

Kittner, Noah, Shabbir H. Gheewala, and Richard M. Kamens. "An Environmental Life Cycle Comparison of Single-Crystalline and Amorphous-Silicon Thin-Film Photovoltaics in Thailand." *Energy for Sustainable Development* 17 (2013): 605–14.

Kollins, Katherine, Bethany Speer, and Karlynn Cory. *Solar PV Project Financing: Regulatory and Legislative Challenges for Third-Party PPA System Owners.* National Renewable Energy Laboratory, NREL/TP-6A2–46723, rev. February 2010. http://www.nrel.gov/docs/.

Lazard. *Lazard's Levelized Cost of Energy Analysis—Version 8.0,* September 2014. http://www.lazard.com/.

Lew, Debra, and Greg Brinkman. *The Western Wind and Solar Integration Study Phase 2.* National Renewable Energy Laboratory, NREL/TP-5500–58798, September 2013. http:// www.nrel.gov/docs/.

Lopez, Anthony, Billy Roberts, Donna Heimiller, Nate Blair, et al. *U.S. Renewable Energy Technical Potentials: A GIS-Based Analysis.* National Renewable Energy Laboratory, NREL/TP-6A20–51946, July 2012. http://www.nrel.gov/docs/.

Lovins, Amory B. *Reinventing Fire: Bold Business Solutions for the New Energy Era.* White River Junction, VT: Chelsea Green, 2011.

Luque, Antonio. "Will We Exceed 50% Efficiency in Photovoltaics?" *Journal of Applied Physics* 110 (2011).

Madaeni, Seyed Hossein, Ramteen Sioshansi, and Paul Denholm. *Comparison of Capacity Value Methods for Photovoltaics in the Western United States.* National Renewable Energy Laboratory, NREL/TP-6A20–54704, July 2012. http://www.nrel.gov/docs/.

Mann, Sander A., Mariska J. De Wild-Scholten, Vasilis M. Fthenakis, Wilfried G. J. H. M. van Sark, et al. "The Energy Payback Time of Advanced Crystalline Silicon PV Modules in 2020: A Prospective Study." *Progress in Photovoltaics: Research and Applications* (2013). http://onlinelibrary.wiley.com/.

Mason, J. E., V. M. Fthenakis, T. Hansen, and H. C. Kim. "Energy Payback and Life-Cycle CO_2 Emissions of the BOS in an Optimized 3.5 MW PV Installation." *Progress in Photovoltaics: Research and Applications* 14 (2006): 179–90.

Maycock, Paul D., and Edward N. Stirewalt. *Photovoltaics: Sunlight to Electricity in One Step.* Andover, MA: Brickhouse, 1981.

Mayfield, Ryan. *Photovoltaic Design & Installation for Dummies.* Hoboken: Wiley, 2010.

Meakin, David H., Cordula Schmid, and Geoffrey S. Kinsey. "Fraunhofer PV Durability Initiative for Solar Modules." *Photovoltaics International* (May 2013): 77–87.

Mehta, Shyam, and Shayle Kann. *The 2014 U.S.–China Solar Trade Dispute: Status, Strategies and Market Impacts.* GTM Research, June 2014.

Mendelsohn, Michael, Travis Lowder, and Brendan Canavan. *Utility-Scale Concentrating Solar Power and Photovoltaics Projects: A Technology and Market Overview.* National Renewable Energy Laboratory, NREL/TP-6A20–51137, April 2012. http://www.nrel.gov/docs/.

Mitchell, Stacy. *How the Walton Family Is Threatening Our Clean Energy Future.* Institute for Local Self-Reliance, October 2014. http://www.ilsr.org.

Moomaw, W., P. Burgherr, G. Heath, M. Lenzen, et al. *Annex II: Methodology,* in *IPCC Special Report on Renewable Energy Sources and Climate Change Mitigation.* Cambridge, UK: Cambridge University Press, 2011. https://ipcc.ch/pdf/special-reports/srren/Annex%20 II%20Methodology.pdf.

Myles, R. W., K. K. Hynes, and I. Forbes. "Photovoltaic Solar Cells: An Overview of State-of-the-Art Cell Development and Environmental Issues." *Progress in Crystal Growth and Characterization of Materials* 51 (2005): 1–42.

National Renewable Energy Laboratory (NREL). *Section 1603 Treasury Grant Expiration: Industry Insight on Financing and Market Implications.* NREL/TP-6A20–53720, June 2012. http://www.nrel.gov/docs/.

Ong, Sean, Clinton Campbell, Paul Denholm, Robert Margolis, et al. *Land-Use Requirements for Solar Power Plants in the United States.* National Renewable Energy Laboratory, NREL/TP-6A20–56290, June 2013. http://www.nrel.gov/docs/.

Pacca, Sergio, Deepak Sivaraman, and Gregory A. Keoleian. "Parameters Affecting the Life Cycle Performance of PV Technologies and Systems." *Energy Policy* 35 (2007): 3316–26.

Peng, Jinqing, Lin Hu, and Hongxing Yang. "Review on Life Cycle Assessment of Energy Payback and Greenhouse Gas Emission of Solar Photovoltaic Systems." *Renewable and Sustainable Energy Reviews* 19 (2013): 255–74.

Perlin, John. *Let It Shine: The 6,000-Year Story of Solar Energy.* Novato, CA: New World Library, 2013.

Pfund, Nancy, and Ben Healey. *What Would Jefferson Do? The Historical Role of Federal Subsidies in Shaping America's Energy Future.* Vote Solar Initiative, August 2011. http:// votesolar.org/.

Prieto, Pedro A., and Charles A. S. Hill. *Spain's Photovoltaic Revolution: The Energy Return on Investment.* Heidelberg and New York: Springer, 2013.

Rocky Mountain Institute. *A Review of Solar PV Benefit & Cost Studies.* April 2013. http://www.rmi.org/.

Rural Community Innovations. *To'Hajiilee Economic Development, Inc.—Utility-Scale Solar Project, Sháńdíín Solar, LLC—Feasibility Analysis.* 2011.

Seel, Joachim, Galen L. Barbose, and Ryan H. Wiser. "An Analysis of Residential PV System Price Differences between the United States and Germany." *Energy Policy* 69 (June 2014): 216–26.

Smith, Sherry L, and Brian Frehner. *Indians & Energy: Exploitation and Opportunity in the American Southwest.* Santa Fe: School for Advanced Research Press, 2010.

Solar Electric Power Association. *Ratemaking, Solar Value and Solar Net Energy Metering—A Primer.* Version 1.0. http://www.solarelectricpower.org/.

Solar Energy Industries Association. *Enlisting the Sun: Powering the U.S. Military with Solar Energy.* May 2013. http://www.seia.org/.

———. *Solar Means Business 2014: Top U.S. Commercial Solar Users.* http://www.seia .org/.

Solar Energy Industries Association and GTM Research. *U.S. Solar Market Insight.* Quarterly and Annual Reports, 2012–2014. http://www.seia.org/ and http://www.green techmedia.com/research/ussmi.

Solar Foundation. *Brighter Future: A Study on Solar in U.S. Schools.* September 2014. http://www.thesolarfoundation.org/.

———. *National Solar Jobs Census 2014: The Annual Review of the U.S. Solar Workforce.* January 2015. http://www.thesolarfoundation.org/.

Terakawa, A. "Review of Thin-Film Silicon Deposition Techniques for High-Efficiency Solar Cells Developed at Panasonic/Sanyo." *Solar Energy Materials & Solar Cells* 119 (2013): 204–8.

Tsosie, Rebecca. "Climate Change, Sustainability, and Globalization: Charting the Future of Indigenous Environmental Self-Determination." *Environmental & Energy Law Policy Journal* 4, no. 2 (Fall 2009).

———. "Keynote Address—Indigenous Peoples and Global Climate Change: Intercultural Models of Climate Equity." *Journal of Environmental Law and Litigation* 25, no. 7 (2010).

US Bureau of Indian Affairs, Bureau of Land Management, Environmental Protection Agency, National Park Service. *Final Environmental Impact Statement: Moapa Solar Energy Center.* February 2014. http://www.moapasolarenergycentereis.com/.

US Bureau of Land Management and Department of Energy. *EIS-0403: Final Programmatic Environmental Impact Statement, Solar Energy Development in Six Southwestern States (AZ, CA, CO, NV, MN, and UT).* July 2012. http://www.energy.gov/.

US Department of Defense, *Annual Energy Management Report, Fiscal Year 2012.* June 2013. http://www.acq.osd.mil/ie/energymgmt_report/FY%202012%20AEMR.pdf.

US Department of Energy. *A Guide to Community Shared Solar: Utility, Private, and Nonprofit Project Development.* May 2012. http://www.nrel.gov/docs/.

——. *Solar Powering Your Community: A Guide for Local Governments.* 2nd ed., January 2011. http://www.solaramericacommunities.energy.gov/.

——. *SunShot Vision Study.* February 2012. http://www1.eere.energy.gov/.

US Department of Energy, Office of Indian Energy. *Developing Clean Energy Projects on Tribal Lands: Data and Resources for Tribes.* Undated. http://www.nrel.gov/, accessed January 1, 2015.

US Environmental Protection Agency, National Renewable Energy Laboratory. Feasibility studies prepared for RE-Powering America's Land Initiative: Siting Renewable Energy on Potentially Contaminated Land and Mine Sites. Multiple dates. http://www.epa.gov/oswercpa/rd_studies.htm.

Warburg, Philip. *Harvest the Wind: America's Journey to Jobs, Energy Independence, and Climate Stability.* Boston: Beacon, 2012, 2013.

Wetzel, Thomas, and Florian Feuerstein. "Update of Energy Payback Time Data for Crystalline Silicon PV Modules." 26th European Photovoltaic Solar Energy Conference and Exhibition, 2011, pp. 3191–195. http://www.eupvsec-proceedings.com/.

Wolfe, Philip. *Solar Photovoltaic Projects in the Mainstream Power Market.* London & New York: Routledge, 2013.

Yue, Dajun, Fengqi You, and Seth B. Darling. "Domestic and Overseas Manufacturing Scenarios of Silicon-Based Photovoltaics: Life Cycle Energy and Environmental Comparative Analysis." *Solar Energy* 105 (2014): 669–78.

Index